TECHNICAL
REPORT

Long Range Energy R&D

A Methodology for Program Development and Evaluation

JAMES T. BARTIS

TR-112-NETL

March 2004

Prepared for the National Energy Technology Laboratory
of the U.S. Department of Energy

The research described in this report was conducted by RAND Science and Technology for the National Energy Technology Laboratory of the U.S. Department of Energy.

Library of Congress Cataloging-in-Publication Data

Bartis, James T., 1945-
Long range energy R&D : a methodology for program development and evaluation / James T. Bartis.
p. cm.
"TR-112."
ISBN 0-8330-3543-6
1. Power resources. 2. Energy development. I.Title.

TJ163.155.A1 B37 2004
333.79'072073—dc22

2003026260

The RAND Corporation is a nonprofit research organization providing objective analysis and effective solutions that address the challenges facing the public and private sectors around the world. RAND's publications do not necessarily reflect the opinions of its research clients and sponsors.

Published 2004 by the RAND Corporation
1700 Main Street, P.O. Box 2138, Santa Monica, CA 90407-2138
1200 South Hayes Street, Arlington, VA 22202-5050
201 North Craig Street, Suite 202, Pittsburgh, PA 15213-1516
RAND URL: http://www.rand.org/
To order RAND documents or to obtain additional information, contact
Distribution Services: Telephone: (310) 451-7002;
Fax: (310) 451-6915; Email: order@rand.org

Preface

This report presents a disciplined planning process for developing and evaluating long-term technology development options within program offices of the U.S. Department of Energy (DOE). This planning process provides a means of articulating long-term needs in energy research and development while accounting for the inability to predict the future, especially the distant future.

This study was commissioned by the National Energy Technology Laboratory as part of its program in support of the Office of Fossil Energy within DOE. The purpose of the study was to develop guidelines for establishing and evaluating performance targets for long-range research directed at advanced fuels and electricity generation.

This report is intended to aid senior-level federal officials responsible for establishing policies and making funding decisions that shape the research and development portfolio within the Office of Fossil Energy. The methodology presented here is also applicable to nonfossil energy technologies and therefore may be of interest to senior-level managers from other DOE offices. Other agencies that support or have cognizance over long-range research and development programs and that could benefit from this report include the U.S. Environmental Protection Agency, the National Science Foundation, the National Institute for Science and Technology, and the White House Office of Science and Technology Policy.

RAND Science and Technology (S&T), a unit of the RAND Corporation, conducts research and analysis that helps government and corporate decisionmakers address opportunities and challenges created by scientific innovation and rapid technology change. The scope of RAND S&T work ranges from emerging energy technologies to global environmental change to other endeavors seeking a better understanding of the nation's scientific enterprise and how best to nurture it. Focal points of RAND S&T work include energy, the environment, information technology, aerospace issues, occupational health and safety, technology and economic development, bioethics, advanced materials, and critical technologies for industries and occupations.

RAND S&T serves a variety of clients, including federal, state, and local government agencies; foreign governments; foundations; and private organizations. RAND S&T researchers have a wide range of expertise. The staff includes physicists and geophysicists; chemists and geochemists; electrical, chemical, mechanical, and information-technology engineers; biological and environmental scientists; and economists and other social scientists.

Inquiries regarding RAND Science and Technology may be directed to:

Stephen Rattien, Director
RAND Science and Technology
1200 South Hayes Street
Arlington, VA 22202-5050
Phone: (703) 413-1100 x5219
Email: contact-st@rand.org
Website: www.rand.org/scitech/

The RAND Corporation Quality Assurance Process

Peer review is an integral part of all RAND research projects. Prior to publication, this document, as with all documents in the RAND technical report series, was subject to a quality assurance process to ensure that the research meets several standards, including the following: The problem is well formulated; the research approach is well designed and well executed; the data and assumptions are sound; the findings are useful and advance knowledge; the implications and recommendations follow logically from the findings and are explained thoroughly; the documentation is accurate, understandable, cogent, and temperate in tone; the research demonstrates understanding of related previous studies; and the research is relevant, objective, independent, and balanced. Peer review is conducted by research professionals who were not members of the project team.

RAND routinely reviews and refines its quality assurance process and also conducts periodic external and internal reviews of the quality of its body of work. For additional details regarding the RAND quality assurance process, visit http://www.rand.org/standards/.

Contents

Preface ... iii
Figures and Tables ... ix
Summary ... xi
Acknowledgments ... xvii
Acronyms ... xix

CHAPTER ONE
Introduction ... 1

CHAPTER TWO
Identifying Long-Range Energy R&D Needs ... 4
Impediments to Long-Range Energy Planning ... 4
Projecting Energy Needs to 2050 ... 5
Purpose and Goals of Long-Term Energy R&D ... 6
Energy Threat Matrix ... 7
Long-Range Energy R&D Goals ... 8
Electric Power ... 9
Processed Liquid Fuels ... 9
Gaseous Fuels ... 10
Relevance and Limitations ... 11

CHAPTER THREE
Evaluating Technologies ... 13
System Evaluation ... 13
Establishing Technology Roles ... 13
Identifying Development Risks and Uncertainties ... 14
Technical Evaluation ... 17
Determining Underlying Fundamental Technical Problems ... 17
Determining R&D Needs and Opportunities ... 17
The IntraTech Approach ... 18

CHAPTER FOUR
Case Study: Power Generation Based on Coal Gasification ... 20
Current Technical Status ... 20
IntraTech Steps 1 and 2: Strategic Analysis ... 22

IntraTech Step 3: Threats and National R&D Needs Addressed by Coal Gasification for Power Generation 22
IntraTech Step 4: Performance and Affordability Risks 23
Carbon Dioxide Concentration and Sequestration 23
High-Efficiency Power Generation..... 24
Risk Summary 26
IntraTech Step 5: Determine Underlying Fundamental Technical Problems 26
Solid Oxide Fuel Cells..... 27
Oxygen-Free Gasification 28
IntraTech Step 6: Determine Research Needs and Opportunities..... 29
Solid Oxide Fuel Cells..... 29
Oxygen-Free Coal Gasification..... 31

CHAPTER FIVE
Conclusions..... 33
Developing High-Level R&D Needs: Steps 1 and 2..... 33
Evaluating Technologies: Steps 3 Through 6 34
Implementing the IntraTech Approach 34

APPENDIX
A. Threat Matrix Analyses 37
B. Current Coal-Fired Power Generation 42

References..... 45

Figures and Tables

Figures

S.1. 2050 Energy Threat Matrix xii
S.2. The Six-Step IntraTech Approach for Long-Range R&D Planning xiv
1.1. The Three Analysis Levels of the IntraTech Approach 2
2.1. 2050 Energy Threat Matrix 8
3.1. The Six-Step IntraTech Approach to Long-Range R&D Planning 19
4.1. Simplified Schematic of a Coal Gasification Combined-Cycle Plant 21
B.1. Effect of Increasing Energy Efficiency on Carbon Dioxide Emissions 44

Tables

S.1. Long-Range Strategic R&D Goals xiii
3.1. Affordability Risk Indicator 15
A.1. Carbon Dioxide Emission Coefficients of Major Fossil Fuels 40

Summary

This report presents a disciplined planning process that links long-range strategic goals to detailed long-term energy research and development (R&D) needs and opportunities. The method begins with an articulation of long-term energy R&D goals, accounting for the difficulty of predicting the far future. It then covers evaluation of the potential of a technology to meet these long-term energy R&D goals, determines the fundamental technical problems that underlie development risks and uncertainties, and identifies alternative development approaches that are directly related to long-term R&D goals.

Most long-term energy planning at the U.S. Department of Energy (DOE) has been based on developing a limited number of projections of future energy supply and demand using large econometric models. Prior projections (made 15 to 25 years ago) of energy use patterns in the year 2000 generally missed the mark by wide margins. Major problems included overestimating energy prices and energy demand. In general, the role of new technologies was overrated, and the evolution of existing technologies was generally underrated. Reviewing more recent energy projections, we conclude that fundamental problems remain. For example, if the Organization of Petroleum Exporting Countries (OPEC) were to lose influence or if highly efficient automotive technologies were to be developed and deployed, world oil supply and demand would change significantly. The approach developed in this report is independent of quantitative projections of energy prices, supply, and demand.

Projecting Energy Needs to 2050

If projecting 25, or even 15, years into the future is so difficult, is it reasonable to attempt to project energy needs nearly 50 years from today? Fortunately, long-term technology development needs can be established without detailed projections of future supply and demand. What we can say with confidence is that 50 years from now, there will be a huge domestic and international demand by end-users for the following types of energy:

- Electric power to meet stationary and some transportation energy needs
- Processed liquid fuels to meet transportation energy needs
- Gaseous fuels to meet stationary and possibly transportation energy needs

For developing a long-term federal R&D portfolio, the amount of any type of energy that may be needed in 2050 is far less relevant than the fact that the amount will be extremely large. For example, even if, contrary to almost all projections of energy demand, demand for electric power drops by as much as 50 percent, a huge amount of electric power will still be needed.

While the U.S. government's energy policies have varied over time, our analysis indicates that the high-level goal of ensuring a steady supply of affordable, environmentally sound energy for America's homes, businesses, and industry has endured for the past 25 years. Consequently, we have adopted this goal and applied it to the long-term portion of the federal energy R&D program. More specifically, the purpose of long-term R&D is to develop technology options that will provide a hedge against future scenarios in which current technologies will no longer be capable of providing steady supplies of affordable electricity and liquid and gaseous fuels.

Long-term threats to adequate supplies of affordable energy fall into the following categories: supply constraints; national security; safety problems; and three environmental issues—air pollution, global climate change, and degradation of land and water. Figure S.1 illustrates how these six problem areas map onto the five main energy sources that currently underlie the U.S. economy.

Strategic long-term energy R&D goals follow directly from the threat matrix and are shown in Table S.1. The prioritization shown in Figure S.1 is neither unique nor static. Future users of this methodology can review problem areas and develop appropriate sets of long-range R&D goals.

The major advantage of the threat-matrix approach, in contrast to scenario-based planning, is that it clearly separates the problem from the solution. This is critical for R&D planning and not easy to achieve with scenario-based approaches, which are generally designed for near-term policymaking rather than long-term R&D planning. Any realistic long-term scenario must include national as well as global responses to problems encountered. If

Figure S.1
2050 Energy Threat Matrix

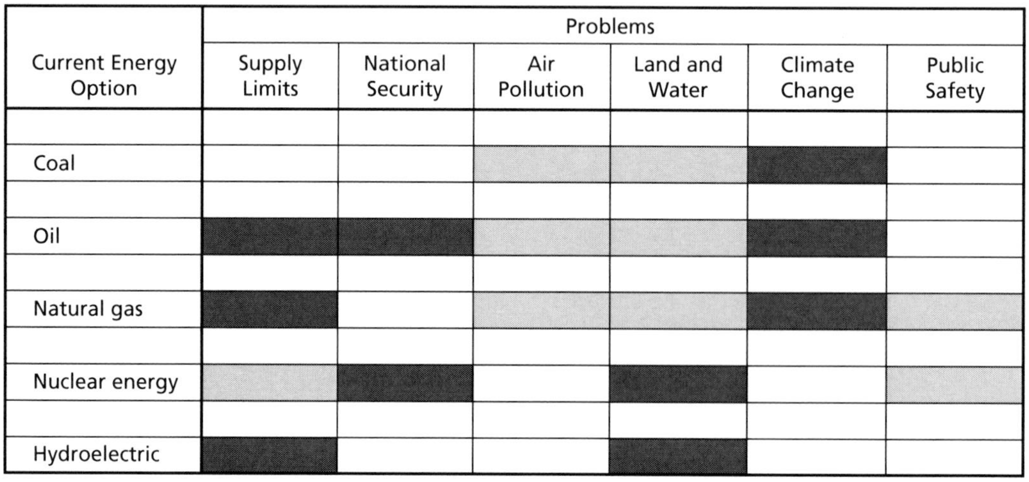

Current Energy Option	Problems					
	Supply Limits	National Security	Air Pollution	Land and Water	Climate Change	Public Safety
Coal						
Oil						
Natural gas						
Nuclear energy						
Hydroelectric						

NOTE: Potential problem areas that threaten the ability to rely on current major energy options in 2050 are indicated in dark gray (most serious) and light gray. The five energy sources listed in the first column provide 97 percent of the energy currently used in the United States.
RAND *TR112-S.1*

Table S.1
Long-Range Strategic R&D Goals

Priority R&D areas are those that offer affordable means of:

For electric power
- Providing electric power
  - — With significantly reduced emissions of greenhouse gases, and
  - — Avoiding nuclear fuel cycle problems, especially proliferation of fission materials and waste fuel disposal
- Reducing demand for electricity from power plants
- Significantly enlarging the technically recoverable resource base for natural gas

For processed liquid fuels
- Meeting transportation energy needs with significantly reduced greenhouse gas emissions
- Reducing energy demand in transportation
- Significantly enlarging the global technically recoverable petroleum resource base
- Providing, with low or no greenhouse gas emissions, non-petroleum sources of liquid fuels

For gaseous fuels
- Increasing supplies of natural gas, including
  - — Enlarging the domestic and global technically recoverable natural gas resource base
  - — Manufacturing synthetic natural gas with low or no greenhouse gas emissions
- Reducing energy demand in traditional and potential new applications for natural gas
- Manufacturing hydrogen with low or no greenhouse gas emissions

such problems and responses are not addressed, the scenarios result in Malthusian projections with, for example, energy becoming very scarce or disastrous climate changes occurring. In some cases, scenarios guess about the costs and performance of technologies that are years away from commercial application. They may also incorporate legislative and regulatory changes that raise ideological controversies not relevant to planning and prioritizing the long-term federal energy R&D program.

The IntraTech Approach

For the purpose of conducting technology-specific analyses, we developed a six-step method, which we call the IntraTech approach. The name reflects its use for linking long-range strategic R&D goals *within* a technology area to detailed, long-term energy R&D needs and opportunities. The IntraTech approach is diagrammed in Figure S.2. The first two steps are directed at understanding the long-term strategic problems that potentially threaten continued reliance on the current predominant energy sources—coal, petroleum, natural gas, nuclear energy, and hydroelectric power. The results of this analysis are the threat matrix and strategic R&D goals described in the previous section.

The next four steps in the IntraTech approach are directed at evaluating a particular technology. The third step consists of determining the specific threats and strategic R&D goals addressed by the technology of interest. This is important because strategic priorities within DOE have evolved since the initiation of development work on many technologies. Additionally, long-term strategic R&D goals are likely to be significantly different from short-term goals.

The fourth step consists of engineering and systems analyses that allow performance uncertainties and affordability risks to be identified. For these analyses, guidelines are provided to establish an even playing field for evaluating affordability risks of advanced energy technologies.

Figure S.2
The Six-Step IntraTech Approach for Long-Range R&D Planning

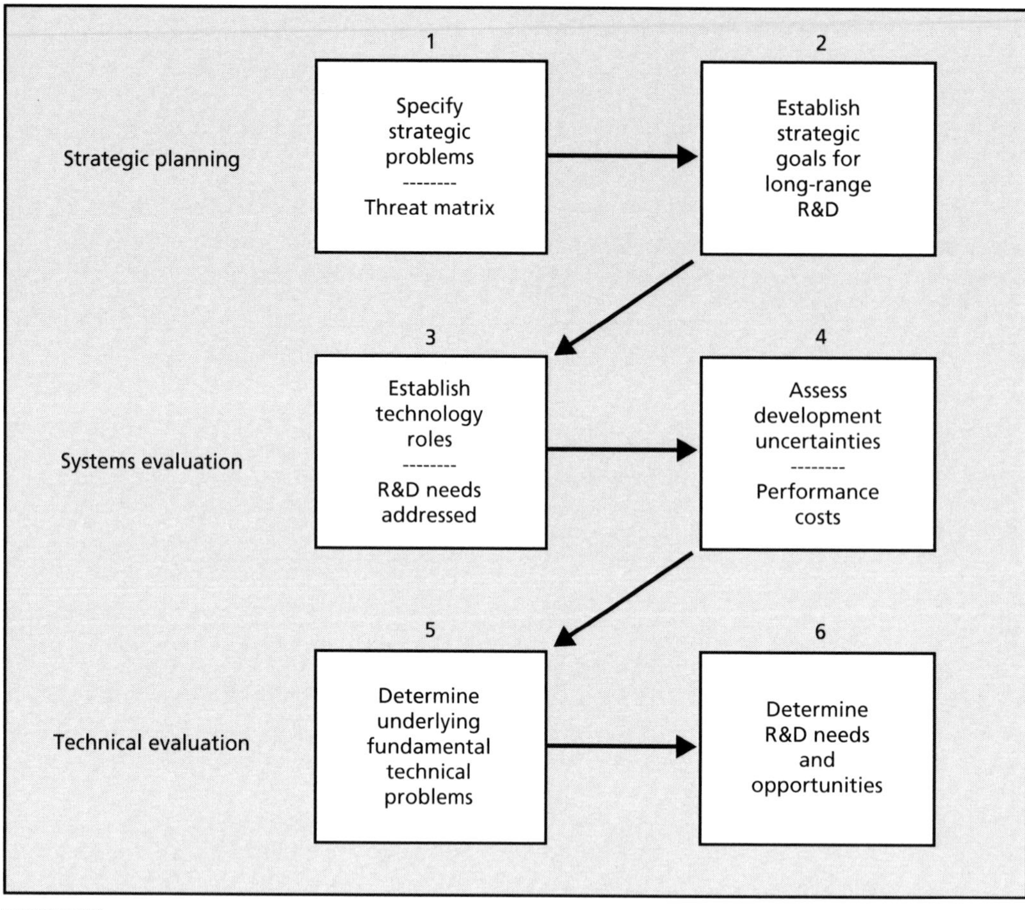

RAND *TR112-S.2*

The fifth and sixth steps consist of technical analyses that determine the fundamental technical problems that underlie performance and cost uncertainties (Step 5) and, using that information, establish R&D needs and opportunities (Step 6).

Case Study: Power Generation Based on Coal Gasification

To illustrate and evaluate the IntraTech approach, we undertook a case study of an ongoing major program area within the DOE Office of Fossil Energy. To limit the scope of this case study, we focused Steps 5 and 6 on two long-term, high-risk areas: low-oxygen gasification and solid oxide fuel cells for central station applications. The case study shows that the IntraTech approach can identify long-term energy goals that a technology concept addresses, identify the key performance factors and risks, identify R&D needs and opportunities, and provide insights regarding alternative development options.

Findings

An advantage, as well as a limitation, of the IntraTech approach is that it stays within a technology area. The method compares the prospective performance and costs of an advanced technology to what is currently commercially available for achieving the desired product. Thus, this approach may be especially useful to DOE officials responsible for planning, justifying, and implementing long-term R&D programs directed at specific technology concepts or groups of related technology concepts.

However, at its current state of development, the IntraTech approach is not appropriate for portfolio management. This limitation became evident in the case study during the analysis of program balance within the DOE solid oxide fuel cell program. Whether the IntraTech "single technology" approach can evolve into a tool suitable for portfolio management remains an open issue. Meanwhile, it can be used to provide senior-level federal officials with uniform information on individual technologies that might better inform the current intuitive/consensus-based process of portfolio management and budget allocation.

Acknowledgments

The author wishes to thank the many members of the RAND staff who provided advice and guidance during the development of the concepts that underlie the planning process and methodologies presented in this report. In particular, Robert L. Hirsch, Paul Holtberg, and Rajkumar Raman played an important role in the initial work leading to the methodology.

Special thanks are extended to Roger Bezdek and his colleagues from Management Information Services, Inc., who conducted a review of prior energy forecasts and provided important insights on the problems and pitfalls involved in trying to develop future scenarios that address long-term energy supply and demand.

Keith Crane, Debra Knopman, and David Bodde reviewed the manuscript and provided important suggestions and valuable advice. An early version of the manuscript benefited from comments received from Jim Dewar.

Greg Kawalkin and Patricia Rawls from the National Energy Technology Laboratory managed this research and provided a constructive critique of the methodology. Raymond J. (Jay) Braitsch, of the Office of Fossil Energy headquarters, provided important guidance and insight.

Acronyms

DOE	U.S. Department of Energy
EIA	Energy Information Administration
HHV	higher heating value
IGCC	integrated gasification combined cycle
LNG	liquefied natural gas
mmBtu	million British thermal units
OPEC	Organization of Petroleum Exporting Countries
R&D	research and development
SECA	Solid State Energy Conversion Alliance
SUV	sport utility vehicle

CHAPTER ONE

Introduction

Since the oil embargo of the winter of 1973–1974, the U.S. government has promoted and supported a national research and development (R&D) program directed at increasing energy availability and improving energy efficiency. This federal program is centered in the U.S. Department of Energy (DOE) and comprises a broad range of energy supply and end-use options. For example, in the electric power generation area alone, the DOE program includes technical concepts using coal, natural gas, nuclear energy, fusion energy, biomass, wind, geothermal energy, and sunlight. Moreover, multiple technical concepts are being pursued for many of these energy inputs. Current DOE spending on energy technology R&D amounts to about $2 billion per year.[1] In many cases, the federal effort leverages additional funding for energy technology development by the private sector.

To manage this large national R&D portfolio, senior-level government officials must make policy and budgetary decisions. The three strategic management issues are

- The balance of the portfolio for addressing near, mid-term, and long-term needs of the nation
- The portfolio's potential for successfully meeting these needs
- The cost-effectiveness of the portfolio versus alternative portfolios

These three issues are also relevant to specific technology development programs within the overall national energy R&D portfolio that are managed by program offices within DOE.

At present, the secretarial level of DOE does not apply an analytical methodology to manage its overall energy R&D program at the portfolio level. This is not a new problem. In 1997, the President's Committee of Advisors on Science and Technology specifically recommended "that the Department make a much more systematic effort in R&D portfolio analysis" (PCAST, 1997). Previously, the Secretary of Energy Advisory Board made a similar recommendation (SEAB, 1995). Departmentwide portfolio management—to the extent that it occurs—emerges from the annual budget-justification process. Programs compete with one another on the basis of analyses conducted by the individual technology-development program offices and the strength of outside stakeholder support.

More systematic management and weighing of technical alternatives generally does occur within the individual technology-development program offices responsible for specific

[1] This includes only the R&D budgets of the following offices within DOE: Office of Fossil Energy; Office of Energy Efficiency and Renewable Energy; Office of Nuclear Energy, Science and Technology; and Office of Fusion Energy Science. Portions of the budgets of other organizations within DOE also support R&D related to energy supply and efficiency. In particular, this estimate excludes R&D related to long-term storage of nuclear wastes.

technologies, such as wind power, photovoltaics, coal liquefaction, coal gasification-based power generation, fuel cells, and carbon sequestration. Engineering analyses, technical assessments, and market assessments are commonly used by these program offices to provide direction for the program managers as well as to defend their programs in the annual budget justification process.

The current method does not include any consistent analytical method for guiding the allocation of limited budget dollars across program lines. Moreover, when assessing the technical and economic hurdles of new technologies, program offices do not apply standardized methods that would enable transparent comparisons across the program areas. Not surprisingly for a budget- and stakeholder-driven process, the current approach tends to overemphasize near- to mid-term payoffs. While this approach has resulted in some notable technical achievements, it is not adequate for long-term strategic management of the nation's energy R&D portfolio. In particular, it is not clear that the parts sum up to a portfolio that is balanced with regard to near- and long-term perspectives.

Achievement of an appropriate balance is impeded by the tendency of government energy R&D program managers to promise early results and to structure their programs accordingly. To a large extent, this is done in response to the political cycle, with federal elected officials tending to have a relatively short time horizon. Another problem is that DOE has given priority to programs and projects that include cost sharing in the form of public-private partnerships. It is not reasonable to expect significant industry cost sharing for R&D directed at long-term energy needs.

In this report, we introduce an alternative planning process that moves a step forward in evaluating technology development options within DOE. We call this process the IntraTech approach. Our focus is on how specific technologies within the national energy R&D portfolio contribute to long-term needs. The methodology underlying this new planning process spans three levels of analyses, as illustrated in Figure 1.1: strategic planning, systems evaluation, and technical evaluation. At the strategic-planning level, the methodology

Figure 1.1
The Three Analysis Levels of the IntraTech Approach

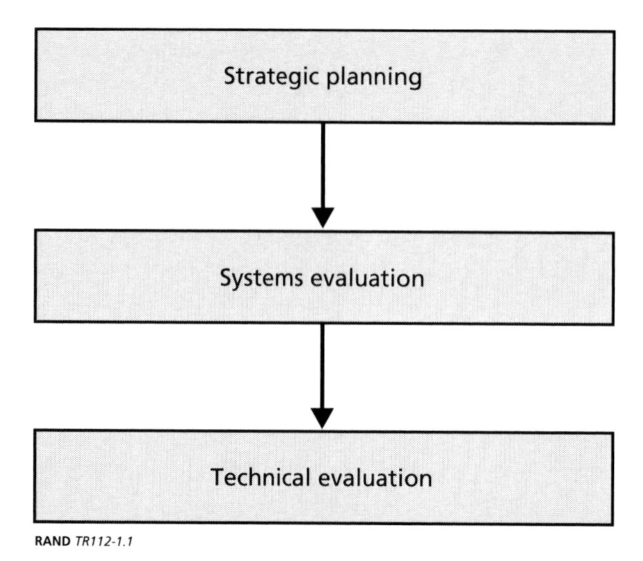

provides a means of articulating long-term needs while accounting for the inability to predict the future, especially the distant future. The strategic-planning component of the methodology is presented in Chapter Two.

Chapter Three presents the systems-evaluation and technical-evaluation components of the methodology. These components build upon similar work currently being performed by the individual technology development program offices. At the systems-evaluation level, the methodology introduces a transparent approach for evaluating the potential of energy systems based on a future technology to meet articulated long-term needs. This is achieved by clearly establishing the technology's potential role in meeting long-term R&D needs and by focusing on the fundamental characteristics of the technology that cause development uncertainties. This component includes establishing cost-estimating guidelines to create an even playing field across the various technology options pursued by program offices within DOE.

At the technical-evaluation level, the methodology establishes a process for identifying long-term R&D opportunities. In this process, a multidisciplinary team of scientists and engineers determines the fundamental technical problems that underlie cost and performance uncertainties. That same multidisciplinary team should then be in a position to determine long-range R&D needs and opportunities.

Chapter Four presents a case study relevant to the Office of Fossil Energy, conducted for the purpose of evaluating the methodology. Selected in consultation with DOE, the case study examines power generation based on coal gasification. Chapter Five summarizes the methodology, evaluates its application in case studies, and discusses its applicability and limitations.

We named this new planning process the IntraTech approach because of its applicability for planning within a technology program area. At its current state of development, the approach cannot be applied to portfolio management. However, the portions of the methodology aimed at identifying long-term needs and development uncertainties should be directly applicable to the more complex problem of portfolio management.

The methodology provides the following information for senior-level technology managers:

1. Long-term energy needs that a technology concept does or could address.
2. Key technical performance factors that favor the technology.
3. Critical technical problems that must be successfully resolved for the technology to become viable.
4. Insight regarding alternative development approaches.

RAND Science and Technology began work on this study by reviewing previous DOE and other studies that have attempted to project future energy supply and demand. This review was undertaken to provide a better understanding of the problems and pitfalls of developing future scenarios that are often used by DOE program managers in formulating and managing R&D programs directed at meeting long-term needs. Under a subcontract with RAND, Management Information Services, Inc., provided assistance in this review. The results of the review are presented in Chapter Two as part of a discussion of impediments to long-term energy planning.

CHAPTER TWO

Identifying Long-Range Energy R&D Needs

Impediments to Long-Range Energy Planning

Long-range energy forecasting is fraught with perils. In this study, 35 long-range energy forecasts were examined to gain insight into the problem of identifying long-term energy technology development needs. We were primarily interested in energy projections and forecasts performed in the 1970s and 1980s. Most of these forecasts or projections were performed using large macroeconomic models that predict how energy supplies and demand will balance, based on a broad range of economic, political, social, and technical assumptions.

How well did these forecasting efforts predict energy supply and use patterns in the year 2000? Our review indicates that they generally missed the mark by wide margins. For example, nearly every study significantly overestimated primary energy consumption in the United States in 2000. Our findings are as follows:

- World energy prices, especially oil prices, are difficult to project; almost all major studies predicted significantly higher prices than were actually experienced in 2000. In general, the longer the time frame, the higher the price projection. In some major studies, projected prices are off by more than a factor of four!
- Energy demand is also very difficult to predict. Forecasters underestimated industrial and consumer decreases in energy consumption in response to the price increases following the Arab oil embargo of 1973 and the Iranian crisis of 1979–1980. Energy forecasters, like everyone else, seem to be incapable of forecasting major changes in technologies, consumer preferences, or external events such as the information revolution, the collapse of the Soviet empire, or the popularity of sport utility vehicles (SUVs).
- The role of entirely new technologies is generally overrated, and the evolution of existing technologies is generally underrated. For example, the future contribution of solar energy has been consistently overestimated. Fusion energy has been 20 to 30 years away for more than 30 years. At the same time, nearly all studies overlooked the role of technical advances in oil and gas exploration and production that have significantly increased supplies and lowered costs.

Even more recent energy projections depend on assumptions with high degrees of uncertainty. For example, the Organization of Petroleum Exporting Countries (OPEC) keeps world oil prices at highly inflated levels. The breakup of this cartel would result in a fundamental change in oil supply and demand. In addition, emerging automotive technologies could significantly reduce demand for refined petroleum products, especially if combined with a consumer shift away from high-weight vehicles. But the timing and extent of such potential reductions in energy demand are highly uncertain, depending on technical, political, and sociological developments.

The value of projections drawn from econometric models is not in their ability to predict the future, but rather in the insights they provide into "what if" policy questions: What if oil prices take a certain trajectory *and* Americans continue to demand SUVs *and* nuclear power is not an option, *and* literally a host of additional assumptions? Such projections are useful for some types of energy-policy analyses, but they can provide a false sense of accuracy and foresight.

Projecting Energy Needs to 2050

If projecting 25, or even 15, years into the future is so difficult, is it reasonable to attempt to project energy needs[1] nearly 50 years hence? Fortunately, detailed projections of future supply and demand are not needed to establish long-term technology development needs. In fact, developing a federal energy R&D portfolio centered on one or a few such projections could result in an overly constrained portfolio that is completely irrelevant to future energy needs.

What can be said with great confidence is that 50 years from today, there will be a huge domestic and international demand by end-users for the following types of energy:[2]

- Electric power to meet stationary and some transportation energy needs
- Processed liquid fuels to meet transportation energy needs
- Gaseous fuels to meet stationary and possibly transportation energy needs

For developing a federal R&D portfolio, the amount of electric power that may be needed in 2050 is far less relevant than the fact that the amount will be extremely large. Current U.S. electric-power consumption amounts to about 3.6 trillion kilowatt-hours (Energy Information Administration, 2003a). This demand is satisfied by nearly 800 gigawatts of generating capacity. Even if, contrary to almost all current projections, electricity consumption drops by 50 percent, the United States will still need a very large amount of electric power.[3] The same logic applies to liquid and gaseous fuels.

[1] The phrase *energy needs* refers to the energy demand to meet transportation needs, domestic heating needs, needs for manufactured products, etc. These fundamental needs drive energy demand.

[2] Coal is not on our list because it is not an important end-use fuel. Today, nearly all coal used in the United States is dedicated to generating electric power.

[3] To put this in perspective, in 1977, about half as much electric power was used in the United States as is used today.

Purpose and Goals of Long-Term Energy R&D

The latest strategic statement of national energy policy puts forth a clear goal for U.S. energy (as distinct from energy R&D) policy: "to ensure a steady supply of affordable energy for America's homes, businesses and industry" (National Energy Policy Development Group, 2001). In September 2003, DOE released a strategic plan covering the department's diverse responsibilities in national defense, energy, nonenergy science, and environmental protection/remediation (U.S. Department of Energy, 2003). With regard to energy, the DOE strategic plan declares that the "principal tool in its national energy policy is science and technology." In fact, the only energy goal (Goal 4) listed in the strategic plan is directed at technology development for the promotion of "energy security," as shown below:

> Improve energy security by developing technologies that foster a diverse supply of reliable, affordable and environmentally sound energy by providing for reliable delivery of energy, guarding against energy emergencies, exploring advanced technologies that make a fundamental improvement in our mix of energy options, and improving energy efficiency. (U.S. Department of Energy, 2003)

The DOE strategic goal expands on the U.S. energy policy goal—a steady supply of affordable energy—by explicitly acknowledging the need for environmental protection. In fact, the high-level goal of ensuring a steady supply of affordable, environmentally sound energy reflects the energy policies of all presidential administrations since DOE's founding in 1977. For example, the 1998 National Energy Strategy (U.S. Department of Energy, 1998) set forth five goals, which together represent a strategic approach for ensuring a steady supply of affordable, environmentally sound energy.[4] Where succeeding administrations have differed has been in the policy approaches and relative emphasis given to promoting energy production, energy efficiency and conservation, environmental protection, and energy research and development.

Because the high-level goal of ensuring a steady supply of affordable, environmentally sound energy has endured for 25 years, it is appropriate to adopt it for the long-term portion of the federal energy R&D program. An example of an enduring, high-level consensus statement adequate for the purposes of pursuing our methodology is shown in the following shaded box:

Purpose of Long-Term Federally Sponsored Energy R&D

Ensure over the long term a steady supply of affordable, environmentally sound energy for America's homes, businesses, and industry

Why ensure? Because there are, on the far horizon (i.e., 50 years from now), a broad range of factors that threaten continued reliance on the way energy supply and demand are currently being met. For example, concerns about global climate change could significantly

[4] The five goals of the 1998 energy strategy were (1) to improve the efficiency of the energy system, (2) to ensure against energy disruptions, (3) to promote energy production and use in ways that respect health and environmental values, (4) to expand future energy choices, and (5) to cooperate internationally on global issues.

alter the way fossil fuels are used. Or geological or geopolitical developments could cause a major increase in world petroleum prices. It is hoped that long-term energy R&D will develop technology options that provide a hedge against future scenarios in which current energy technologies will no longer be able to provide the amounts of electricity, liquid fuels, and gaseous fuels that are likely to be needed.

The National Energy Policy statement does not define *affordable*. By definition, energy is affordable if its costs can be borne without serious detriment. In this report, we adopt a slightly more stringent definition for the purposes of R&D planning, namely, that the direct *and* indirect costs of energy[5] should not constitute a significantly larger fraction of average household income than they currently do. For the purposes of developing a planning methodology, we define a significant increase as one that is greater than 25 percent.

The purpose of federally sponsored long-term energy R&D does not explicitly encompass reducing the dollar cost of energy supplies. Reduction of dollar costs is a very weak goal for federally sponsored research and one that many would argue belongs to the private sector. At present, through a combination of domestic production and imports, the United States is able to meet all of its energy needs at fairly reasonable prices. When energy prices become unreasonably high, the cause is usually a temporary energy infrastructure failure or actions of OPEC to control world oil prices. The methodology developed in this report does not address the goal of reducing energy costs below current levels.

Energy Threat Matrix

Long-term, i.e., 50-year, threats to adequate supplies of affordable energy fall into the following categories: supply constraints; national security considerations; safety problems; and three environmental issues—air pollution, global climate change, and degradation of land and water. Figure 2.1 illustrates the level of threat that these six problem areas pose to the five main energy sources in the current U.S. economy. This mapping is based on an analysis of the various problem areas over the long term, presented in Appendix A, but it is not unique. Future users of this methodology can review these problem areas, using the latest available information, and can develop their own threat matrices.

The threat matrix is not a prediction or a forecast of the future; rather it shows problem areas that are likely to persist and that might seriously limit future use of particular energy options that are important today. Not all threats are equally likely to materialize or to have equal impact. For example, some might argue that air pollution considerations are highly unlikely to constrain fossil fuel use given recent trends in environmental control systems and clean-fuel technology. The review in Appendix A indicates that climate change, oil and natural gas supply limits, national security, and nuclear-fission fuel-cycle issues appear as high-impact threats. They are designated in dark gray in Figure 2.1. The lack of economic and environmentally acceptable sites seriously constrains increased reliance on hydropower.

[5] Consumers pay for energy directly through purchases of electricity and fuels and indirectly through purchases of products whose production costs include energy.

Figure 2.1
2050 Energy Threat Matrix

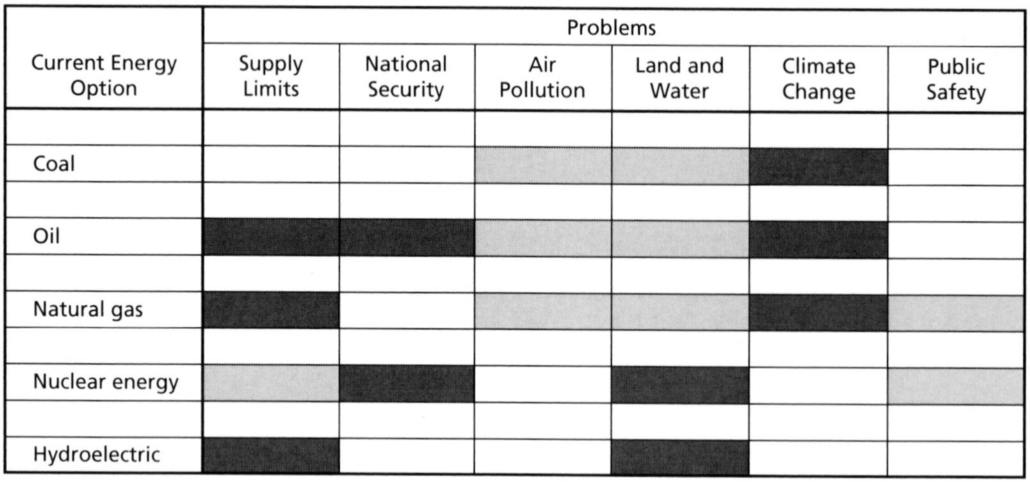

Current Energy Option	Problems					
	Supply Limits	National Security	Air Pollution	Land and Water	Climate Change	Public Safety
Coal						
Oil						
Natural gas						
Nuclear energy						
Hydroelectric						

NOTE: Potential problem areas that threaten the ability to rely on current major energy options in 2050 are indicated in dark gray (most serious) and light gray. The five energy sources listed in the first column provide 97 percent of the energy currently used in the United States.
RAND *TR112-2.1*

The threat matrix pertains to 2050. A matrix addressing the nearer-term situation might have a different set of threats or might emphasize different threats. For example, a near-term threat matrix would almost surely give greater emphasis to air pollution and safety of liquefied natural gas (LNG) terminals.

We do not list renewable or essentially inexhaustible energy sources, such as wind, photovoltaic power, geothermal energy, or biomass-derived fuels as current options in the threat matrix. At their present state of development, such options do not offer an affordable means of meeting a significant share of national energy demand. Rather, we have taken the approach that long-term R&D (and maybe nearer-term R&D) directed at renewable and inexhaustible options has the potential to offer solutions that transcend the persistent threats to continued reliance on current energy sources and technologies. Alternatively, such advanced technologies could be included in a separate threat matrix covering options that do not currently meet the affordability criteria.

Long-Range Energy R&D Goals

General statements of R&D priorities can be developed directly from the energy threat matrix. Basically, a high-priority R&D area is one that offers to eliminate or significantly attenuate one or more of the most serious threats (the dark gray boxes) to continued supplies of electric power, liquid fuels, and gaseous fuels.

Electric Power

Power plants using coal, nuclear energy, and natural gas provide more than 85 percent of the power generated annually in the United States.[6] Over the long term, continued use of all three energy sources could be severely limited by potential problems, as discussed in Appendix A: climate change in the case of coal and, to a lesser extent, natural gas; supply limits in the case of natural gas; and fuel-cycle problems (nuclear-weapons proliferation and waste disposal) in the case of nuclear energy. Lack of economic and environmentally appropriate sites limits significant growth of hydroelectric power generation.

A number of technical alternatives have been identified to address these problems. For example, greenhouse gas emissions can be reduced by using renewable technologies such as wind power and photovoltaic power cells, by generating electric power at higher energy efficiencies, by decreasing power losses in electricity transmission and distribution, and/or by capturing and permanently sequestering carbon dioxide. Additional options include increasing the use of nuclear power and/or shifting from coal to natural gas to take advantage of the lower carbon dioxide coefficient of natural gas (Appendix A). But turning to these latter two options requires that other problems be addressed successfully, namely, the nuclear fuel-cycle problems of proliferation and waste disposal and limited supplies of natural gas. All of the foregoing options address the supply side of electric power, but options that cause decreased demand for electricity from power plants are also effective. Demand-side options, such as increasing the efficiency of electrical appliances, are especially attractive, since they offer to lessen the magnitude of all the problems identified in the threat matrix. Consideration of the above technology options in the context of the energy threat matrix leads to the long-range R&D goals summarized in the following shaded box:

Electric Power

Priority areas for long-range R&D are those offering affordable means of

- Providing electric power
  - —With significantly reduced emissions of greenhouse gases, and
  - —Avoiding nuclear fuel-cycle problems, especially proliferation of fission materials and waste fuel disposal
- Reducing demand for electricity from power plants
- Significantly enlarging the technically recoverable resource base for natural gas

Processed Liquid Fuels

Refined petroleum products currently supply 99.7 percent of the energy used by the transportation sector, excluding natural gas used to operate pipelines. Over the long term, use of petroleum as the principal fuel for transportation could be limited by depletion of reserves, national security issues, and possible constraints on emissions of greenhouse gases. The long-range R&D goals that derive from the threat matrix are shown in the shaded box below:

[6] Of the remaining 15 percent, hydropower supplies about 10 percent; petroleum combustion supplies about 3 percent; and wood and waste (municipal, industrial, and agricultural) combustion supplies about 2 percent.

Processed Liquid Fuels

Priority areas for long-range R&D are those offering affordable means of

- Reducing energy demand in transportation
- Meeting transportation energy needs with significantly reduced greenhouse gas emissions
- Significantly enlarging the global technically recoverable resource base for petroleum
- Providing non-petroleum sources of liquid fuels with low or no greenhouse gas emissions

As with electric power, advanced technologies that could reduce energy demand in transportation address all of the potential threats to liquid-fuel supplies shown in the energy threat matrix. The second bullet specifically addresses advanced technology options that offer to reduce greenhouse gas emissions in transportation, such as hydrogen-fueled or electric vehicles. The third bullet addresses the problem of geographically limited petroleum reserves, and the fourth bullet recognizes the potential opportunities to expand the resource base for liquid fuels through advanced technology directed at liquids from renewable resources, as well as from fossil fuel resources such as oil shale, natural gas, and coal, provided that greenhouse gas emissions can be appropriately controlled.

Gaseous Fuels

Natural gas is an important fuel in the residential, commercial, and industrial sectors. Over the past dozen years, natural gas use in electricity generation has doubled. Power generation now accounts for about 23 percent of the annual natural gas consumption in the United States. Since natural gas has the lowest carbon-to-hydrogen ratio of any fossil fuel, it releases the least carbon dioxide per unit of energy produced through combustion.

The critical long-term problems for gaseous fuels, as shown in the energy threat matrix, are limited resources of natural gas and carbon dioxide emissions. Technology advances that reduce the use of natural gas address both of these problems. Supply limits may be addressed by development of advanced technology directed at increasing natural gas supplies, for example, by extracting gas from the vast amount of methane hydrates found in the polar regions of North America and the deep-water portions of the continental shelf. Alternatively, natural gas can be manufactured from renewable resources such as biomass and other fossil fuel resources such as coal, but long-range R&D for increasing supplies should focus on technical approaches that produce low emissions of greenhouse gases.

Hydrogen is a gaseous fuel that has the advantage of not emitting any carbon dioxide. Hydrogen is not a naturally occurring fuel, but must be manufactured using natural gas (currently the method of choice), other fossil fuels, or electricity. Other methods are being researched. These considerations support the priority long-range R&D areas as listed in the shaded box below:

Gaseous Fuels

Priority areas for long-range R&D are those offering affordable means of

- Reducing energy demand in traditional and potential new applications for natural gas
- Increasing supplies of natural gas, including
  - Enlarging the domestic and global technically recoverable resource base for natural gas
  - Manufacturing synthetic natural gas with low or no greenhouse gas emissions
- Manufacturing hydrogen with low or no greenhouse gas emissions

Relevance and Limitations

Our 2050 threat matrix shows four major problem areas that severely threaten long-term reliance on energy options: global climate change limiting fossil fuel use; national security problems associated with increased reliance on nuclear power and petroleum; fuel-cycle problems limiting reliance on nuclear power; and supply problems limiting long-term availability of petroleum and natural gas as well as expansion of hydroelectric power. Each of the current energy options faces a serious threat. It is this fact that underlies the establishment and mission of DOE and the importance of its energy research programs.

The major advantage of the threat matrix approach is that it clearly separates the problem from the solution. This is critical for R&D planning, and it is not easy to achieve using scenario-based approaches designed for near-term policymaking rather than long-term R&D planning. Any realistic long-term scenario must include national as well as global responses to problems encountered. Otherwise, the scenarios result in Malthusian projections, such as energy becoming very scarce or potentially disastrous climate changes occurring. However, when solutions are included in scenarios, the scenarios themselves become controversial. In some cases, they embed guesses about the costs and performance of technologies that are years away from commercial application.

Additionally, scenarios are often used to reflect projected legislative and regulatory changes. For example, a new regulatory structure, both international and domestic, will be required if global emissions of greenhouse gases are to be controlled. The ideological controversies raised by details of alternative regulatory structures are not relevant to planning and prioritizing the long-term federal energy R&D program.

The threat matrix approach has been designed for goal-oriented government-sponsored technology development programs. Thus, it may be applicable to other federal R&D programs, including those undertaken within the Department of Defense, Department of Homeland Security, Department of Commerce, Department of Agriculture, Department of Health and Human Services, and the Environmental Protection Agency. But at its current state of development, the threat matrix approach is not applicable to private sector

R&D. Unlike governments, private firms are driven by profits, and this provides them with a unifying measure for evaluating alternative options.

Not all government-sponsored R&D is goal-oriented technology development. Basic and fundamental research programs where the objectives are to advance progress in science and promote scientific and technical education are not amenable to the threat matrix approach. This is not to say that there is a well-defined line between basic and fundamental research and goal-oriented technology development. Gaps in understanding fundamental processes can seriously impede long-term technology development, as discussed in the case study presented in Chapter Four.

The threat matrix and the strategic R&D goals developed in this chapter can be used to order and evaluate individual energy technology programs directed at long-term energy problems. But at its current state of development, our methodology lacks the driver for R&D planning that dominates scheduling and budget. That key driver is the required or desired timing of successfully meeting a strategic goal. Consider the 2050 priority R&D areas for electric power. Here, the key threats are global climate change, limited supplies of natural gas, nuclear-weapons proliferation, and nuclear waste fuel disposal. The resolution of each of these problems requires a schedule. At present, DOE does not appear to have an analytic framework that can be used to drive the pace and budgets of its R&D programs.

For example, scheduling and budgeting long-term electric power generation R&D to address global climate change requires resolving the following questions: Is it adequate and necessary to have an electric power generating technology base available (but not necessarily deployed) by 2050 that offers an affordable means of reducing greenhouse gas emissions 80 percent from current levels? Is 80 percent enough, or is it too much? Is 2050 too early or too late? Similar questions pertain to natural gas supplies and nuclear fuel-cycle problems.

Because we have not addressed these issues analytically, the methodology developed here cannot be used to drive budgets and program scheduling. Progress in this area will require careful analyses that incorporate risks and uncertainties.[7] It is likely that such analyses will require the development of scenarios that capture the potential environmental or fuel-supply consequences of alternative schedules of meeting strategic R&D goals.

---

[7] For example, RAND's Frederick S. Pardee Center for Longer Range Global Policy and the Future Human Condition is currently examining the timing of decisionmaking regarding global climate change.

CHAPTER THREE

Evaluating Technologies

This chapter presents the system evaluation and technical evaluation components of the IntraTech approach. It is important to understand which goals a particular technology effort addresses, especially for portfolio analyses. The principal emphasis of this chapter is on identifying (1) the technical risks that might prevent successful development and (2) the kind and type of R&D that is most appropriate. This is a very difficult problem because advocates and stakeholders of various energy R&D efforts are generally not forthcoming regarding development risks.

System Evaluation

Establishing Technology Roles

The system-evaluation component of the methodology is relatively straightforward, assessing how various energy systems that incorporate an advanced technology address the energy threat matrix and the long-range R&D goals developed in Chapter Two. For example, successful development of high-temperature fuel cells might enable affordable distributed generation and cogeneration systems that operate at exceptionally high energy efficiencies. For such systems based on natural gas, this component of the methodology would quantify the potential role of the technology in addressing the long-term threats to continued natural gas use as shown in the energy threat matrix. Basically, higher efficiency operations (including reduced transmission losses) would allow a moderate reduction in carbon dioxide emissions per unit of delivered electric power, compared with current state-of-the-art, central station combined-cycle power plants fired with natural gas. A more significant impact on greenhouse gas emissions would result if the development of high-temperature fuel cells led to lower-cost approaches for capturing carbon dioxide for eventual sequestration. In this case, the technology would clearly address the long-term R&D goal of *providing electric power with significantly reduced emissions of greenhouse gases.*

In some instances, an advanced system of interest may be supplied with an energy source not listed in the energy threat matrix. In these cases, it is important to understand potential issues that could seriously limit the contribution of the advanced technology. Alternative energy sources may raise new threats or may have inherent technical limitations. For example, until affordable means of storing energy are developed, wind and photovoltaic power systems will not be able to displace significant amounts of conventional generating capacity, because wind and solar energy are naturally variable, and conventional generating capacity will be required to assure power on demand.

Identifying Development Risks and Uncertainties

Strategic planning of energy R&D requires a means of identifying development risks and uncertainties and communicating them to high-level decisionmakers. This can be accomplished by using performance and affordability risk indicators.

The performance risk indicator identifies major unresolved performance or operational problems that are critical to the technology's success in meeting important energy R&D needs. Three examples illustrating this performance risk indicator are presented in the shaded box below. In cases where the knowledge base is not sufficient to support a conceptual design and cost analysis, the technology development effort should be designated as involving high risks.

Nearly all of the energy technology concepts in the DOE R&D portfolio have been under development for some time, in many cases more than 20 years. This affords senior-level R&D managers an opportunity to evaluate R&D programs against past promises. Where progress is lacking, the program should be reevaluated to determine whether development challenges are being underestimated. If they are, development risks should be reassessed and program emphasis should be reconsidered, with a possibly significant shift from near-term to longer-term research.

Examples of Performance/Operational Risks

1. Carbon sequestration: It is unclear whether either deep-aquifer or ocean sequestration of carbon is technically feasible and environmentally acceptable. For deep-aquifer sequestration, the major uncertainties concern the ability of the aquifers to accept high flows of carbon dioxide and the ultimate fate of carbon dioxide within the aquifer. For ocean sequestration, the key uncertainty is the impact of large flows of carbon dioxide on ocean flora and fauna.

2. Magnetic fusion power generation: All currently available structural materials become highly radioactive after relatively brief exposure to neutrons from a deuterium-tritium burning plasma. The fusion research community recognizes that a "low-activation" material must be developed to avoid a massive radioactive waste disposal problem. A few candidate systems have been investigated, but significant additional R&D is required to resolve confidently this uncertainty.

3. Hydrogen cogeneration from solid-state fuel cells: Engineering analyses show that high-temperature solid-state fuel cells offer a low-cost, high-efficiency approach to generating hydrogen. Basically, all required fuel cell cooling is performed by the endothermic reaction that produces hydrogen from hydrocarbons. But at present adequate experimental information is not available to support an evaluation of the technical feasibility of the hydrogen cogeneration concept.

For many technologies in development, performance or operational problems translate into affordability issues. For example, the state of the art in electricity storage systems significantly limits the use of renewable technologies that are by nature intermittent, such as photovoltaic power systems and wind turbines. But batteries, albeit expensive, are commercially available, so the storage problem eventually comes down to costs.

Cost estimates for advanced systems are often based on the current development status of the enabling technologies. When such estimates are available, they can be compared to costs of current commercial systems that meet similar energy needs. The ratio of these two costs provides a useful indicator of the technical risks associated with developing an affordable alternative. A three-value example is illustrated in Table 3.1. It is reasonable to expect that decisionmakers and planners may select other cutoffs for the cost ratio, depending on the nature of the overall portfolio and the planning environment. The use of new-to-current cost ratios requires a disciplined process of engineering analyses and costing, as presented in the cost evaluation guidelines discussed in the following sections.

Table 3.1
Affordability Risk Indicator
(an example of how cost ratios can be related to the technical risks of successfully developing an affordable technology)

Cost Ratio (new/current)	Affordability Risk Indicator
< 1.5	Low
1.5 — 5	Moderate
> 5	High

The Competitive Environment. Let us assume that we wish to evaluate the potential of a technology to enter the market within the next five to ten years. In this case, the technology should be fairly well defined, although uncertainties regarding costs and reliability sometimes persist until late in the development process (Merrow, Phillips, and Myers, 1981). The competitive environment—i.e., the entire suite of marketplace factors that determine whether the emerging technology can successfully compete—should also be fairly well defined. These marketplace factors include the following:

- Competing technologies
- Tax laws
- Regulation and license requirements: environmental, safety, and permitting
- Prices of primary fuels
- Economic climate, e.g., inflation, interest rates, and availability of venture capital

An evaluation of the near-term potential of a technology is meaningless unless it considers the competitive environment, because that near-term potential is determined by whether, and to what extent, the technology can enter the marketplace and successfully compete with current and other emerging technologies.

Moving to the far future, the competitive environment becomes highly uncertain. The difficulty of defining this environment is a key impediment to long-range energy planning, as discussed in Chapter Two. Hundreds of energy technologies are under development and which of these will be successful is highly uncertain. To assume that tax laws affecting energy choices will remain unchanged over the next 40 to 50 years is also without foundation. Tax laws will change because fiscal policies offer powerful means of addressing potentially critical energy issues, such as reducing greenhouse gas emissions, promoting energy conservation, and reducing dependence on imported fuels. Taxes, subsidies, and certain types of regulations are used to promote energy choices, including technology choices, that are viewed as offering social benefits but that would not be made in the absence of government action. Such fiscal tools, however, merely shift costs from one group of energy users to another or to the overall tax base.

Fortunately, evaluation of a long-range R&D portfolio does not require a projection of the competitive environment in the far future. In fact, trying to include such a projection can lead to serious misjudgments regarding the economic and social benefits of a successful development of an advanced technology.

Cost Development Guidelines. The following guidelines can be used to establish an even playing field for evaluating affordability risks of advanced energy technologies for long-term applications. Since the far future is highly uncertain and the details of the competitive environment 50 years hence are irrelevant, the objective underlying these guidelines is to understand the bare costs associated with an advanced energy technology, unburdened by taxes and free of subsidies.

- Determine the costs of an advanced energy system based upon the current state of the art of the technical concept being evaluated. For large field-erected systems, assume an "nth plant" design. For small manufactured systems, assume assembly-line production. To avoid clouding a technology's development status, make no assumptions regarding learning-curve effects.
- Avoid comparative analyses of prospective technologies. Instead, compare costs and performance to those of existing commercial systems that yield the same or similar products. In particular, define system boundaries to meet required energy demand. For example, evaluations of technologies that produce intermittent power, such as photovoltaic and wind turbine systems, need to include backup systems required to produce power on demand. End-use systems for higher-efficiency energy use should be evaluated against saved fuel costs.
- Conduct all technology evaluation analyses on a pre-tax basis.
- Ignore assumptions about the general rate of inflation by using real or constant-dollar values in economic evaluations.
- Allow primary fuel prices to vary over a broad range of values. Engineering analyses often allow fuel prices to escalate over the life of the cost analysis, but this practice can cause deceptive results and therefore is not recommended.
- Conduct comparative analyses using a low and a high real rate of return; do not assume debt financing.

Underlying these guidelines is the assumption that when it comes to a long-term R&D portfolio, the government is generally not in a position to forecast which technologies will be the economic winners and thus systems of choice. Presumably, successful pursuit of a long-term R&D program for a particular technology concept would eventually enable the private sector to make investment decisions regarding near-term development and marketing.

Technical Evaluation

Once development risks and uncertainties have been identified, a combination of engineering and scientific analyses can determine the fundamental technical problems that drive development risks and underlie uncertainties. This knowledge provides the foundation for determining R&D needs and opportunities, the final step in the IntraTech approach to long-term R&D program planning.[1]

Determining Underlying Fundamental Technical Problems

The purpose of this step is to dig further down into fundamental scientific and technical issues that underlie the assessment of performance and operational risks or affordability risks. Where performance and operational risks have been identified, in-depth analyses are required of the fundamental technical problems or issues that must be addressed to remove uncertainties. Where affordability risks have been determined for a technical concept, engineering analysis of the conceptual design can determine the critical technical problems that are causing the high costs. For example, the key cost driver for photovoltaic power systems is system efficiency. The technical problems underlying low system efficiency are associated with the power-conversion efficiency of the photovoltaic cell, as well as losses in other parts of the system, such as resistance losses in wiring and terminals and losses in power conditioning. Each of these key components can be further examined to identify the fundamental scientific phenomena that limit performance of near-, mid-, and long-term options.

Identifying fundamental problems that underlie the technical risks of success in the long-term portion of the R&D portfolio is not a simple, straightforward task. For example, analysis of the costs of a coal gasification system shows that the gasifiers represent a very small portion of the capital costs of a commercial plant. But it would be a major mistake to assume that long-term R&D on coal gasification should ignore the gasifier, since its design drives requirements for oxygen production, steam production, heat exchange, and downstream processing.

Determining R&D Needs and Opportunities

The final step in the IntraTech approach consists of determining R&D needs and opportunities. Since we are addressing goal-oriented, long-term programs or program components, our focus is on R&D to solve fundamental technical problems or reduce major uncertainties. In-depth scientific knowledge, imagination, and nonlinear and interdisciplinary thinking are needed to develop the R&D needs and opportunities for such a program. Returning to the photovoltaic system as an example, one of the major cost components of such a system is the structure required to support the cells. It took imagination to recognize that using photovoltaic cells as a roofing material could be a solution to this problem.

---

[1] Planning near-term and mid-term R&D programs requires an additional step, namely, determining the appropriate U.S. government role, given development activities of the private sector and other nations.

Scientists and engineers tend toward specialization, and this can be problematic when evaluating risks and identifying R&D needs and opportunities. For example, in many cases materials scientists are not directly involved in assessing, planning, developing, and managing energy technology development programs. As a consequence, materials research is often neglected as a solution for addressing technical problems in energy technologies directed at long-range energy needs. To overcome the problem of specialization, the development of R&D needs and opportunities should involve teams of scientists and engineers covering a broad range of disciplines.

The IntraTech Approach

We recommend a six-step process for identifying R&D needs for meeting long-term energy needs, as illustrated in Figure 3.1:

1. Develop a threat matrix and prioritize its cells. The threat matrix presented in Chapter Two can serve as a starting point. The threat matrix should be common to all technologies being evaluated.
2. Using the threat matrix, determine the long-term energy R&D goals that, if achieved, would eliminate the serious threats in the matrix. The R&D goals listed in Chapter Two can serve as a starting point.
3. For a specific technology, determine the threat matrix cells and the energy R&D goals that the technology could address. Be cognizant of new threats or problems that the new technology might introduce.
4. Determine performance and/or affordability risks associated with each R&D goal. Affordability risks should be determined using the cost-development guidelines presented.
5. Determine the technical and scientific problems that drive risks.
6. Determine and evaluate technology-specific R&D needs and opportunities.

Steps 1 and 2 were developed in Chapter Two. Step 3 is straightforward. For Step 4, it is important to recognize that a single technology could be low-risk for some threat/R&D goals and high-risk for others. For example, indirect liquefaction of coal is probably a low-risk technology area for addressing the oil-supply threat and the energy R&D need to "provide non-petroleum sources of liquid fuels." On the other hand, indirect liquefaction may be a very high-risk approach for "meeting transportation energy needs with significantly reduced greenhouse gas emissions."

Steps 5 and 6 are related in that they require creative analyses by high-caliber scientists and engineers from all relevant disciplines, including materials research.

On completion of these six steps, program managers will have the essential ingredients they need to include long-term research needs and opportunities within program plans.

Figure 3.1
The Six-Step IntraTech Approach to Long-Range R&D Planning

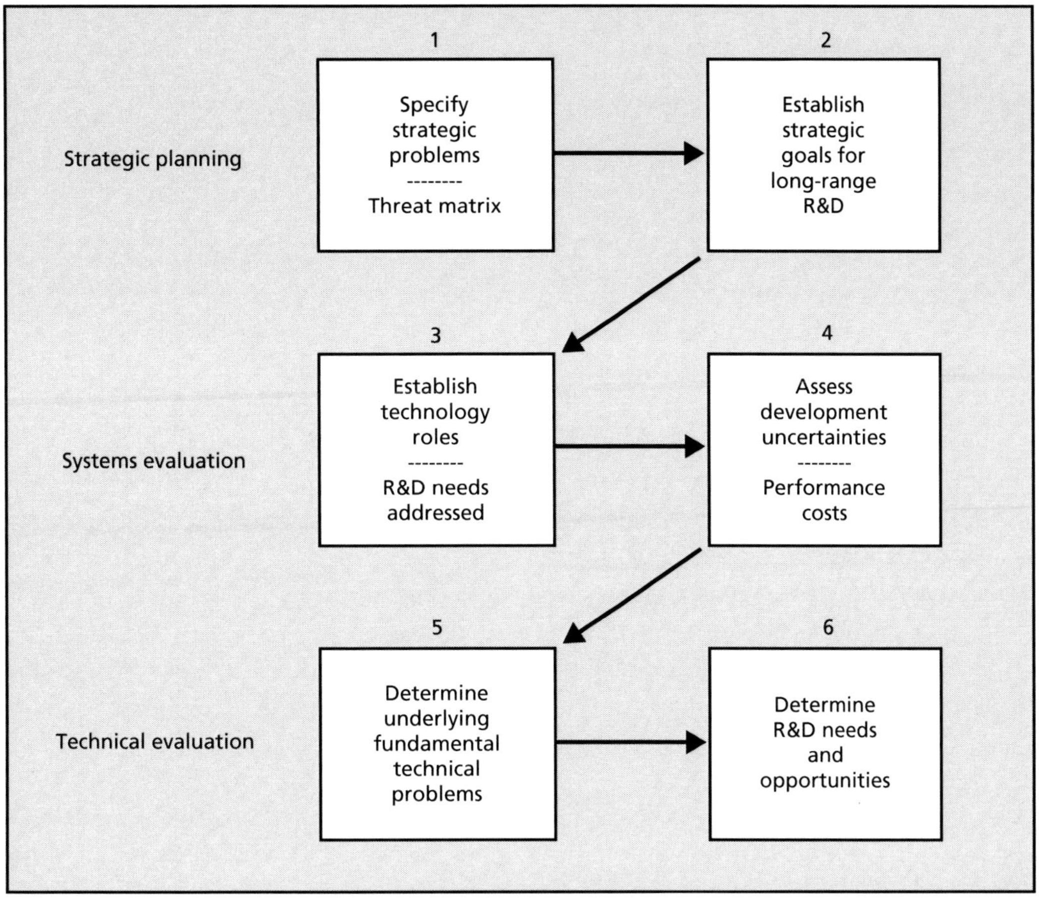

Strategic planning

1
Specify strategic problems

Threat matrix

2
Establish strategic goals for long-range R&D

Systems evaluation

3
Establish technology roles

R&D needs addressed

4
Assess development uncertainties

Performance costs

Technical evaluation

5
Determine underlying fundamental technical problems

6
Determine R&D needs and opportunities

RAND *TR112-3.1*

CHAPTER FOUR

Case Study: Power Generation Based on Coal Gasification

For more than 30 years, DOE and its predecessor agencies have sponsored research to develop power generation methods based on coal gasification. The principal drivers of this effort have been reducing the cost of coal use, reducing air emissions, and increasing energy efficiency, all as compared to the conventional method of generating power from coal, namely, pulverized coal combustion.[1]

Beginning in the late nineteenth century, coal gasification plants became common in both Europe and the United States. Coal-derived gas was the "town gas" used for municipal lighting and distributed to homes and businesses for lighting and cooking. In the early 1950s, the extension of natural gas pipelines into New England brought about the closing of the last U.S. coal gasification plants.

In most coal gasification power-generation concepts, coal is reacted with oxygen and steam to produce a combustible mixture of gases consisting primarily of carbon monoxide and hydrogen. This mixture is often called *synthesis gas*, because it can be used to synthesize a variety of chemical compounds.

Current concepts for using coal gasification to generate power involve deeply cleaning the synthesis gas and sending it to either a gas turbine or an assembly of high-temperature fuel cells. The high-temperature gases leaving the gas turbine or the fuel cell would have sufficient energy to generate additional power. In the case of the gas turbine system, this energy would be used to raise steam that is sent to a steam turbine/generator system, as shown in Figure 4.1. This dual-cycle approach allows higher energy efficiencies, as discussed below.

In system concepts using high-temperature fuel cells, the exit gases can be sent to a gas turbine and then to a heat exchanger, where steam would be raised. This triple-cycle system offers the opportunity for even higher efficiencies. Because power-generation systems based on coal gasification generally involve multiple generation cycles, the technology is commonly termed integrated gasification combined cycle (IGCC).

Current Technical Status

Under the DOE Clean Coal Technology Program, the federal government shared in the cost of two projects that successfully demonstrated IGCC power generation at full commercial

[1] Readers who are not familiar with conventional coal-fired power generation should consult Appendix B.

Figure 4.1
Simplified Schematic of a Coal Gasification Combined-Cycle Plant

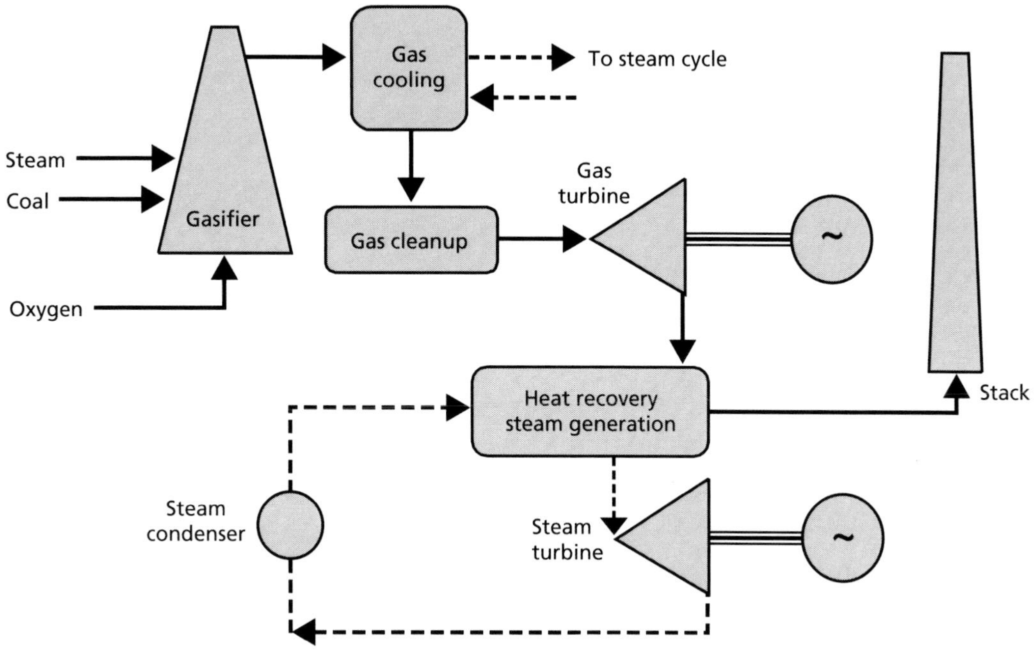

NOTE: The solid lines represent air/gas flow, and the dashed lines represent steam flow.
RAND *TR112-4.1*

scale. Both demonstrations began operation in the mid-1990s. Although the demonstrations are contractually completed, both projects continue to operate as commercial power plants.

At the current state of the art, IGCC generation produces less air pollution than does the conventional approach (pulverized-coal combustion) for generating electric power from coal. However, this is a minor benefit with regard to controlling sulfur oxide, nitrogen oxide, and particulate emissions, since environmental control systems for pulverized-coal combustion power plants are capable of very high levels of pollutant reduction. More important, IGCC plants can use control technology that has demonstrated between 90 and 95 percent reductions in mercury emissions (Ratafia-Brown et al., 2002a). It is uncertain whether such high levels of mercury reduction can be reliably achieved with pulverized-coal plants.

Energy efficiencies of nearly 40 percent[2] have been obtained in the first-of-a-kind demonstration plants. At the present state of the art, both IGCC and pulverized-coal combustion plants can probably demonstrate energy efficiencies as high as 42 percent, but such high-efficiency plants are unlikely in the United States, since low coal prices do not motivate the extra investment required for high-efficiency power generation. The average coal-fired power plant in the United States operates at an energy efficiency of 33 percent, and newer coal-fired plants have a design efficiency of about 36 percent.

Another benefit of IGCC power plants is a general reduction in solid wastes. In particular, the sulfur compounds removed in the gas cleanup system are converted to saleable products, either pure sulfur or sulfuric acid. For most gasifiers, the coal ash is rejected as a

---

[2] All energy efficiencies reported in this report are based on the higher heating value (HHV) of the fuel.

highly nonleachable slag that can be used in manufacturing operations such as cement production.

The near-term market prospects for IGCC are uncertain. Over the past 15 years, coal has not competed favorably with natural gas as a fuel for new power plants. But natural gas prices appear to be increasing, and when and where the price of delivered natural gas approaches $5.00 per mmBtu, coal becomes a competitive option.

The second issue clouding the near-term market prospects for IGCC is the competition from conventional pulverized-coal combustion. Under the current federal regulatory framework for environmental protection, the additional environmental benefits offered by IGCC do not offset a cost advantage of about 20 percent that is held by conventional pulverized-coal power plants. Pending legislation, i.e., the Clean Skies Act of 2003,[3] calls for caps on emissions of sulfur and nitrogen oxides and limits on mercury emissions. Passage of this legislation would improve IGCC's competitive posture, but pulverized-coal combustion may still remain the system of economic choice. Market entry for IGCC may require cost reductions from targeted near-term technology development.

IntraTech Steps 1 and 2: Strategic Analysis

The first two steps of the IntraTech approach consist of strategic analyses that are not technology-specific. Consequently, we can use the threat matrix (Step 1) and the goals for long-term R&D (Step 2) developed in Chapter Two as the strategic foundation for the case study.

IntraTech Step 3: Threats and National R&D Needs Addressed by Coal Gasification for Power Generation

The coal row in the threat matrix lists three long-term problems for continued use of coal for electricity generation: air pollution, degradation of land and water, and global climate change. Currently available IGCC technology significantly addresses air pollution associated with coal use. In addition, solid-waste disposal and/or storage problems at an IGCC power plant site are significantly mitigated by the production of recyclable, possibly saleable, by-products and the transformation of the coal ash to a highly nonleaching slag.

In the near term, power-generation plants based on coal gasification should be able to attain operating efficiencies of 42 percent. Plants operating at this efficiency use about 14 percent less coal than those operating at 36 percent, which is typical of today's newer coal plants. Fourteen percent less coal use translates into a 14 percent decrease in carbon dioxide emissions and a commensurate reduction in land disturbed by mining. However, addressing global climate change will likely require much more than a 14 percent reduction in carbon dioxide emissions.

---

[3] The Clean Skies Act of 2003, an Administration initiative, was introduced in the U.S. House of Representatives as House Bill HR 999 and the U.S. Senate as Senate Bill S. 485.

DOE-sponsored engineering studies (Buchanan et al., 2002; Ratafia-Brown et al., 2002a, 2002b; Ruth, 2002) show that successful long-term R&D on IGCC power generation would offer the following two benefits:

1. Low-cost means of concentrating carbon dioxide for sequestration. Depending on the approach used, between 85 and 99 percent carbon capture is possible.
2. Energy efficiencies of between 50 and 60 percent. This range corresponds to a 28 to 40 percent reduction in coal use and carbon dioxide emissions from the levels associated with operating at an energy efficiency of 36 percent.

Long-term R&D on power generation based on coal gasification addresses all three threats to continued coal use, including the most serious problem, that of global climate change. With regard to long-range energy R&D needs, successful development of advanced IGCC technology would offer a *means of providing electric power with significantly reduced emissions of greenhouse gases, while avoiding nuclear fuel-cycle problems.*

IntraTech Step 4: Performance and Affordability Risks

Carbon Dioxide Concentration and Sequestration

About 1 billion tons of coal are mined each year in the United States. Roughly 70 percent of the weight of coal is carbon. On combustion, this coal releases into the atmosphere about 2.5 billion tons of carbon dioxide. A single coal-fired power plant typically releases between 400 and 800 tons of carbon dioxide per hour.

The concept behind carbon dioxide concentration and sequestration is to capture carbon dioxide at the power plant site and transport it to a location where it can be sequestered and thereby prevented from entering the atmosphere. First, a concentrated, high-pressure stream of carbon dioxide must be obtained. For today's pulverized-coal plants, this is an expensive process, because the flue gas leaving the plant has a carbon dioxide content of about 15 percent by volume, its major constituent being nitrogen. Moreover, the flue gas is at atmospheric pressure. Once the carbon dioxide is captured, it must be compressed to between 20 and 80 atmospheres, the higher pressure allowing pipeline transport as a supercritical fluid. Such high levels of compression are both expensive and energy-consuming. Engineering analyses indicate that carbon dioxide removal and pressurization costs would increase the cost of power generated at a state-of-the-art coal-fired power plant based upon pulverized-coal combustion by 50 to 60 percent (Buchanan et al., 2002). Additionally, energy efficiency would drop from 40 percent to 29 percent, necessitating a 38 percent increase in coal mining to meet the same electric power demand.[4]

In contrast, the combustible gas stream leaving an oxygen-blown coal gasifier is not diluted with nitrogen and is already pressurized to about 20 atmospheres. After cleaning, commercially available processes can convert this pressurized stream to a mixture in which hydrogen and carbon dioxide are the primary constituents. Typically, the carbon dioxide content of this stream would be about 40 to 45 percent by volume. Engineering analyses of a

[4] This and the following paragraph address costs and energy losses associated with carbon dioxide capture and compression. Carbon dioxide transport and sequestration will incur additional costs and energy losses.

slightly advanced IGCC plant show that 90 percent carbon removal would decrease plant energy efficiency from 43.1 percent to 35.4 percent and would increase electricity production costs by about 30 percent (Buchanan et al., 2002). Long-term R&D directed specifically at reducing the costs of carbon dioxide collection and compression is unlikely to have a significant impact. Far greater opportunities are promised by advanced cycles that offer higher-efficiency power production at lower costs, as discussed below.

Transportation costs for pressurized carbon dioxide should be roughly similar to those of long-distance pipeline transport of natural gas. The impact on net power generating costs will depend on the required transport distance—for distances of a few hundred miles, a 5 to 10 percent increase appears reasonable.

All concepts that involve sequestering massive amounts of carbon dioxide present major technical uncertainties. The technical feasibility and environmental impacts of the two major options—deep ocean sequestration and geological sequestration—have not been established. Because of these technical uncertainties, timetables and firm goals for resolving technical, environmental, and economic issues have yet to be established.

Of the three steps involved in carbon dioxide sequestration, the weak link dominating performance and economic uncertainties is the final sequestering. Risks and uncertainties can be summarized as follows:

- Capture and concentration. Presents low affordability risk (Table 3.1 of Chapter Three), but seriously reduces power plant energy efficiency, which requires additional coal production, resulting in adverse impacts to land and water due to increased mining and solid-waste disposal requirements. Increased coal use also produces more carbon dioxide requiring sequestration.
- Transport. Impacts on costs and energy efficiency are small; long-term R&D is unlikely to offer significant benefits.
- Sequestering. Dominates performance and economic uncertainties. Until uncertainties associated with sequestering massive amounts of carbon dioxide are resolved, the combination of carbon sequestration and coal gasification-based power generation should be viewed as a high-risk option.

High-Efficiency Power Generation

As part of its Vision 21 program, the DOE Office of Fossil Energy and the associated National Energy Technology Laboratory have identified a number of advanced power generation concepts that offer electricity production (without carbon sequestration) at energy efficiencies of between 50 and 60 percent. As mentioned earlier, increasing efficiency from 36 percent to these levels would reduce coal use and carbon dioxide emissions by 28 and 40 percent, respectively.[5]

Depending on the resolution of scientific, economic, and socio-political issues involved in addressing the threat of global climate change, a 40 percent reduction in carbon dioxide emissions at power plants, combined with advanced technologies for more efficient end-use of electricity, may be an adequate and appropriate response. For example, a 50 per-

[5] An energy efficiency of 36 percent is typical of coal-fired power plants recently constructed in the United States. Compared to the average efficiency of all coal-fired power plants, 33 percent, advanced plants operating at efficiencies of 50 to 60 percent would reduce coal use and carbon dioxide emissions by 34 to 45 percent.

cent reduction in end-use demand for power combined with a 40 percent reduction in emissions at power plant sites would reduce carbon dioxide emissions from coal-fired electric power generation by 70 percent.

Should deeper levels of control be required, carbon sequestration would be necessary if coal is to continue as a major fuel for electricity generation. In this case, developing higher-efficiency means of coal gasification-based electricity generation would also yield significant benefits. For example, adding carbon dioxide collection and compression systems to a plant with 60 percent efficiency would decrease its efficiency to 52 percent, in contrast to the 34 percent efficiency that would likely result from adding a carbon dioxide removal system to an IGCC plant that would otherwise operate at the current state-of-the-art efficiency of 42 percent. The difference between the advanced plant and existing IGCC technology would be a 35 percent reduction in both coal mining and the amount of carbon dioxide requiring capture, compression, transport, and sequestration.

Advanced technology concepts for high-efficiency power production generally involve one or more of the following (Ruth, 2002):

1. Oxygen production at lower cost and reduced energy consumption
2. Improved gas-gas separation technology
3. Higher-temperature gas purification systems
4. Higher-efficiency gas turbines
5. Higher-temperature steam cycle components
6. High-temperature fuel cells
7. Reduced or eliminated oxygen demand during gasification

The first five areas involve developing improved versions of current technology as well as taking radical departures from current practice. The former is appropriate for nearer-term R&D and is generally associated with low to moderate risks. In particular, successful pursuit of near-term research may boost IGCC energy efficiency up to 45 percent and reduce capital and operating costs sufficiently for IGCC to become the system of economic choice for coal-based power generation.

Attaining economic power generation systems with operating efficiencies of 50 to 55 percent will require a major technological breakthrough in one of the seven areas, or significant technical advances in multiple areas. For example, oxygen production currently accounts for 15 to 25 percent of the cost of operating an IGCC plant and consumes 15 to 20 percent of the plant's power output (Ruth, 2002). Major advances in membranes for separating oxygen from air could result in significant cost and energy savings (Area 1), but those savings would need to be accompanied by progress in other areas to achieve operating energy efficiencies of more than 50 percent. A potentially higher payoff is possible by completely eliminating the need for oxygen during gasification (Area 7). This higher-risk/higher-payoff approach is further developed below.

Since research directed at 50 to 55 percent energy efficiency can proceed in seven distinct areas, each of which has multiple options, a robust R&D plan can be developed to pursue this objective. While the performance risk for any one approach may be high, the risk for the "portfolio" is moderate.

Risks become much higher if the goal for advanced IGCC systems is raised to energy efficiencies of 60 percent or greater. Attaining such high efficiency will require the successful development of economic high-temperature, pressurized fuel cell technology suitable for application in central station power plants, or a fundamental breakthrough in oxygen-free coal gasification. In either case, significant advances would also be required in one or more of the other six areas listed above.

Engineering prototypes of high-temperature fuel cell systems have been demonstrated at the multihundred-kilowatt level, using natural gas as a fuel. The demonstrated technologies include solid oxide fuel cell systems that operate at temperatures of 900°C to 1000°C and molten-carbonate fuel cell systems that operate at 650°C to 700°C. Because of the current high costs of these demonstration systems and the further requirements associated with interfacing in an IGCC power plant, high risks accompany the development of high-temperature fuel cells for multi-cycle coal gasification-based power generation. With regard to oxygen-free gasification, the development risks are also high. Over the past 20 years, little research effort has been devoted to the more complex gasification technology required for oxygen-free gasification.

Reaching the goal of 60 percent energy efficiency for IGCC power systems requires a development path with at least one major technological breakthrough, in either fuel cells or oxygen-free gasification. Consequently, reaching 60 percent or more in energy efficiency should be viewed as a high-risk goal.

Risk Summary

The development of advanced IGCC technology offers the potential of *producing electric power with significantly reduced emissions of greenhouse gases.* If deep reductions (i.e., greater than 40 percent) of carbon dioxide emissions are required at coal-fired generating plants, the critical uncertainty is the ability to sequester massive amounts (billions of tons per year) of carbon dioxide. Costs and environmental impacts of implementing a sequestration strategy would be greatly reduced by the development of IGCC technology that achieves energy efficiencies above 50 percent at reasonable costs.

Beyond enabling economic carbon dioxide sequestration, high-efficiency IGCC could go a long way toward reducing carbon dioxide emissions, especially when combined with end-use energy conservation measures. Raising energy efficiency to between 50 and 55 percent is considered a moderate-risk goal, because a relatively large number of technical paths can be pursued. Reaching 60 percent or greater energy efficiency will require the successful development of either high-temperature fuel cells for central power applications or oxygen-free coal gasification systems.

IntraTech Step 5: Determine Underlying Fundamental Technical Problems

Step 5 of the IntraTech methodology in our case study addresses the technical and scientific problems associated with developing high-temperature fuel cells for IGCC applications and oxygen-free coal gasification. For high-temperature fuel cells, our case-study focus will be on systems involving solid oxide cells. Our analysis of oxygen-free gasification is more cursory, since very little experimental data are available. For both options, attention is directed at the long-term portion of the R&D portfolio.

As noted earlier, we wish to illustrate how the IntraTech approach can identify long-term research opportunities that can have a major impact on resolving major threats to continued energy use. As mentioned at the conclusion of Chapter Three, implementing Steps 5 and 6 requires creative analyses by high-caliber scientists and engineers covering all relevant disciplines, including materials research. For the purpose of this case study, however, this instruction is not being followed, since we are attempting only to illustrate the implementation of the IntraTech approach rather than to perform a definitive analysis of technical problems and long-term research needs and opportunities. The limitation of our discussion to solid oxide fuel cells and oxygen-free gasification is not based on a judgment that other research avenues are less interesting or less productive; rather these two options provide compact illustrations of how the IntraTech methodology can illuminate the importance of research areas that are currently not viewed as mainstream and are not significantly funded, within the context of IGCC development.

Solid Oxide Fuel Cells

Fuel cells are similar to electric batteries in that both convert chemical energy into a combination of electrical energy and heat energy. In a battery, the reacting chemicals are packaged within the device; when they run low, the battery is unable to generate power. In contrast, the reacting chemicals are continuously added to a fuel cell. As long as the reactants flow, the fuel cell can generate power. Acceptable fuels for solid oxide fuel cells include the key components of the synthesis gas from a coal gasifier—hydrogen and carbon monoxide. Methane is also an acceptable fuel.

A fuel cell subsystem for a central station power plant would typically consist of tens of thousands of fuel cells. A typical fuel cell operates at between 0.6 and 0.9 volts. To achieve higher voltages, fuel cells are placed in stacks, with the positive terminal of each cell contacting the negative terminal of its neighbor. The fuel cell subsystem includes provisions for fuel and air delivery to and heat removal from each cell within each stack.

The technical problems faced in developing economic, highly efficient solid oxide fuel cell subsystems for IGCC operations primarily concern cell performance, stack performance and assembly, and systems integration (Singhal and Dokiya, 1999; Yokokawa and Singhal, 2001). For cell performance, the key problems concern energy efficiency, power density (watts/cm^2 of cell surface area), and thermal and chemical degradation of cell materials. A poor understanding of the fundamental electrochemical phenomena underlying cell performance impedes progress in this area.[6]

Major technical problems in stack performance are associated with stack power density, energy-conversion efficiency, fuel and air distribution, fuel utilization, and waste-heat management. Approaches to all these problems must allow economic assembly of the thousands of stacks that would likely be required in a central station application.

In general, systems integration dominates near-term R&D focused on bringing an advanced technology to the marketplace. But systems integration also plays an important role in defining long-term R&D directions. For example, in large central station power-generation concepts involving IGCC and solid oxide fuel cells, waste-heat management is a critical problem driving efficiency and costs. At the current state of the art, about 50 percent of the chemical energy delivered to a fuel cell stack is converted to heat, which must be removed to prevent overheating. Improving fuel cell and stack energy efficiencies serves to lessen this problem, but even highly successful efforts in raising energy efficiency would not eliminate the need for extensive heat management.

---

[6] Personal communications from fuel cell developers and members of the DOE fuel cell program management team.

A second systems-integration problem concerns fuel utilization. In a solid oxide fuel cell, fuel is delivered to the anode, where it reacts with oxygen ions that moved from the cathode via the solid oxide electrolyte. On reaction, water vapor and carbon dioxide are generated at the surface of the anode and mix with the incoming fuel. This mixing causes fuel to be lost as the water vapor and carbon dioxide leave the anode surface. Obtaining satisfactory fuel cell stack performance at fuel utilization levels above 80 percent is currently a major challenge for fuel cell developers. Raising fuel utilization increases electricity generating efficiency but reduces overall power density, affects corrosion, and leads to larger and more expensive stacks.

A number of promising ways to address this problem exist at the component development level. There are also interesting system approaches. For example, in IGCC/fuel cell system concepts, the unused fuel could be used to raise the temperature of the gas exiting the fuel cell subsystem before it is sent to a gas turbine. While the unused fuel is put to good purpose, however, it basically misses the fuel cell, causing a net loss in system efficiency.

Oxygen-Free Gasification

The basic chemical reaction in coal gasification is between coal and water, producing synthesis gas (carbon monoxide and hydrogen):

$$C(\text{coal}) + H_2O \rightarrow CO + H_2 ,$$

where C represents the carbon content of the coal. In entrained-flow gasifier systems, such as those of Texaco, Shell, and Destec, which are designed to produce synthesis gas, this reaction occurs at high temperatures, 1100°C to 1500°C. This reaction is endothermic, i.e., it absorbs heat. For every mole of carbon, the reaction requires 32 kilocalories of heat.[7] That heat is supplied by introducing oxygen into the gasifier, where additional coal is burned, producing carbon dioxide and some additional carbon monoxide. Oxygen, rather than air, is used to prevent dilution by nitrogen, which would cause an approximately threefold increase in the volume of the product gas.

A fundamental problem is how to introduce heat into a gasifier without using oxygen. This is especially difficult because materials for heat transfer are subject to thermochemical degradation at the high temperatures at which gasification normally occurs. Another approach is to convert coal directly to methane (CH_4) within the gasifier, according to the following reaction:

$$2C(\text{coal}) + 2H_2O \rightarrow CH_4 + CO_2 .$$

This reaction is nearly energy-neutral, requiring only one kilocalorie of heat per mole of carbon. In addition, it significantly decreases the volume of the gas stream exiting the gasifier, thereby reducing downstream gas processing costs. The energy efficiency and potential cost advantages of going directly to methane have been recognized for decades. For example, Exxon Research and Engineering, with support from DOE, pursued a catalytic coal gasification process that was based on the observation that coal catalyzed with potassium salts promotes methanation of coal gasification products. Work on this concept was begun in the

[7] Based on the heat of reaction of graphite at 800°C.

early 1970s but ceased in the mid-1980s, apparently in response to falling energy prices. During the course of developing this process, Exxon researchers identified technical problems with catalyst recovery, corrosion of materials, and reducing carbon losses in solid-waste streams.

During the 1970s, DOE funded a few other advanced gasification processes directed at maximizing methane yield. Most of these processes involved transferring hot solids at high pressures and ran into technical difficulties, especially when continuous operations were attempted during process development. These processing problems, combined with a significant decrease in domestic natural gas prices, caused the DOE coal gasification program to shift its emphasis away from methane production and toward the high-temperature coal gasifiers that are appropriate for synthesis gas production.

IntraTech Step 6: Determine Research Needs and Opportunities

Step 6 of the IntraTech approach consists of the identification and analysis of research needs. Because this step requires inputs from technical experts covering a broad range of disciplines and technologies, the following discussion is intended only to illustrate how the IntraTech approach can connect high-level objectives (Steps 1 and 2) to research needs and opportunities. As in the previous step, the case study is generally limited to long-term research needs associated with solid oxide fuel cells; oxygen-free coal gasification, another high-risk area, is discussed briefly.

Before proceeding with the case study, we must emphasize the importance of research directed at understanding the environmental and fundamental technical issues associated with sequestering massive amounts of carbon dioxide. In addition, a robust long-term research program is needed to successfully develop IGCC systems that operate at energy efficiencies of between 50 and 55 percent. This goal is moderately risky, rather than highly risky, only because of the richness of possible approaches to attaining it. Premature narrowing of research options would convert this moderate-risk option to a high-risk option.

Solid Oxide Fuel Cells

Technical issues and research needs are well-documented by the solid oxide fuel cell research community (Singhal and Dokiya, 1999; Yokokawa and Singhal, 2001) and the DOE program office responsible for developing solid oxide fuel cell technology (National Energy Technology Laboratory, 2002). The following discussion, therefore, will address two strategic R&D issues regarding long-term progress: (1) basic and applied research needs and (2) the balance among market opportunities.

Basic and Applied Research. Discussions with members of the research community and a review of the literature indicate a need for long-term basic and applied research directed at the following:

- Understanding fundamental electrocatalytic phenomena and the processes that affect current flow and energy losses in solid-state fuel cells.
- Developing and exploring the properties of new material sets that offer improved energy efficiency, power density, and/or performance at lower temperatures.

Because of a combination of budget constraints and pressure to obtain tangible near-term results, the DOE technology development program supports very little research in these areas. A review of the DOE Basic Energy Sciences program and those of the National Science Foundation also revealed that very little federal support is directed to understanding fundamental electrochemical phenomena relevant to solid oxide-type electrolytes.[8]

Lack of federal support for basic and applied research on solid oxide fuel cells is a major programmatic flaw that could seriously impede the long-term potential of such fuel cells. While recent progress in solid oxide fuel cell development is encouraging, significant performance improvements are required to reduce costs and maintain high-efficiency performance.

Developing new material sets offers a major opportunity to reduce costs and improve performance. Exploratory research has already identified new material sets that offer solid oxide fuel cell operations at much lower temperatures (i.e., below 600°C) than are currently possible. Finding the right set of materials is a time-consuming and complicated process; researchers must consider not only improved performance, but also reliability, raw materials costs, and manufacturing potential. Work is also needed to develop material sets appropriate for improved high-temperature performance, costs, and reliability. In particular, a combination of high- and low-temperature fuel cell stacks could significantly improve waste-heat management and the performance of IGCC fuel cell systems.

Program Balance. To make a major difference in how energy is used, solid oxide fuel cells need to penetrate one or more of the following very large markets:[9]

- Motive power for vehicles (about 50 kilowatts for automobiles, about 1 megawatt for large trucks)
- Distributed generation of electric power using a clean fuel such as natural gas (250 kilowatts to 5 megawatts)
- Centralized generation of electric power using coal (50 to 300 megawatts)

There are also numerous markets that are not significant from an energy perspective but that may provide early opportunities for introducing solid oxide fuel cell technology into commercial use. Customers in these markets may value the unique performance aspects of solid oxide fuel cell systems highly enough to be willing to pay the premium associated with a new technology.

The DOE program for solid oxide fuel cell development is centered on the Solid State Energy Conversion Alliance (SECA) plan to produce affordable 3- to 10-kilowatt solid oxide fuel cell modules that would be powered by natural gas and used in mobile and specialty stationary applications.[10] SECA is a near-term program that aims for early products suitable for niche markets within a few years. High energy efficiency in power generation is not an important SECA program goal; rather, the primary benefit of success would be to

---

[8] This review was conducted using the RAND RaDIUS database as part of an ongoing RAND project examining strategic options for high-temperature fuel cell development.

[9] The required fuel cell system generating capacity is shown in parentheses.

[10] Examples of markets appropriate to SECA goals are isolated homes, ranches and farms, auxiliary power systems for trucks, auxiliary and/or portable power for military operations, and backup power systems in environmentally stressed areas.

speed the initial market entry of solid oxide fuel cell technology, thereby motivating greater private sector investment in technology that could eventually offer high-efficiency power generation appropriate to one or more of the three big markets listed above.

In addition to SECA, DOE is supporting limited work on highly efficient (i.e., about 56 percent) distributed generation technology using solid oxide fuel cells in combination with a gas turbine. Nearly all of this work is directed at near-term payoffs; emphasis is on technical concepts that could be ready for initial demonstration within about five years.

With a limited budget, restricting the DOE program to small-scale systems and early markets appears to be a reasonable decision. But the SECA program does not address major problems associated with larger decentralized or centralized systems for power generation, e.g., heat management and fuel utilization. The strategic issue is whether the DOE program should be enhanced to support R&D relevant to the large potential markets.

Of the three large markets, the obvious candidate for an expanded program is distributed generation. As developers of low-temperature fuel cells are discovering, performance and cost requirements of the automotive market are exceptionally demanding. Therefore, stationary applications are being seriously examined as potential routes for early market entry of commercial low-temperature fuel cell systems. Moreover, the technology base for stack and component development that would be demonstrated in a successful natural gas-based distributed generation program would be highly applicable to centralized IGCC power generation using fuel cells.

The primary R&D needs for fuel cell applications in IGCC systems are low-cost stack designs, reliable materials, fuel utilization, and heat management. All of these issues can be addressed within a technology development program directed at distributed generation systems using natural gas. Because distributed generation systems are a factor of 100 to 1,000 smaller than centralized systems, development and demonstration costs would be considerably less. A program directed at centralized IGCC applications makes sense only in the context of successfully pursuing a solid oxide fuel cell system appropriate for distributed generation using natural gas or some other clean fuel.

Addressing program balance requires resolution of a budget issue: Should solid oxide technology development be funded to cover a new program directed at distributed generation options? Resolving this issue requires that DOE establish schedules for meeting its strategic R&D goals and, in the context of such schedules, review its total portfolio of energy R&D to determine whether a reprioritization is appropriate.

Oxygen-Free Coal Gasification

Most analyses of advanced IGCC systems involve oxygen-blown, high-temperature coal gasifiers in which methane is not a product. In general, lower gasification temperatures and higher hydrogen pressures cause the gasifier yield to shift from synthesis gas toward methane. Since the energy need of the gasifier is determined by the production of synthesis gas, high methane yields imply reduced requirements for oxygen.

The Exxon catalytic coal gasification process is one of a number of low-oxygen or oxygen-free gasification approaches that may be appropriate for advanced IGCC applications. Other examples include the "advanced gasification-combustion" approach that involves three coupled fluidized bed reactors and hydrogasification. The former is in pilot-scale development (Ruth, 2002). Hydrogasification, the reaction of high-pressure hydrogen and coal to produce methane in the absence of steam or oxygen, has been subjected to a scoping analysis which showed very high efficiency when coupled with a high-temperature fuel cell system. However, a more definitive analysis requires experimental investigation of reaction rates and yields.[11]

Most of the work on low-oxygen and oxygen-free gasification concepts was terminated more than 25 years ago. Since then, performance objectives and requirements have changed significantly. No longer is the principal goal the production of coal-derived methane as a substitute for natural gas. High-efficiency electric power generation and carbon dioxide capture are much higher priorities today than they were in the 1970s, when limited supplies of petroleum and natural gas dominated energy policy.

Moreover, recent and prospective advances in a number of enabling technologies open new paths for oxygen-free gasification. Considerable progress has been achieved on advanced membrane systems capable of separating gases such as hydrogen or methane at high or at least very warm temperatures. Although high-temperature gas cleanup has proved elusive, considerable progress appears possible in developing warm, as opposed to cryogenic, systems for removing sulfur and other impurities from the gasifier exit gas. Finally, the potential benefits of methane for managing the waste heat generated by a solid oxide fuel cell give added value to a gasifier exit stream with high methane content. Beyond these specific technical advances, general advances have been made in high-temperature materials, instrumentation, and solids transfer.

Developing an economic approach to oxygen-free coal gasification is a high-risk endeavor. The experience of the 1970s was disappointing, as low-cost methods of producing a synthetic natural gas from coal were not successful. But the issues that now drive long-term energy research—in particular, global climate change—make it advisable to revisit this approach. Moreover, advances in enabling technologies offer a potentially rich opportunity for long-term research directed at oxygen-free gasification of coal for IGCC applications.

---

[11] Personal communication with Rodney A. Greisbrecht, National Energy Technology Laboratory, April 14, 2003.

CHAPTER FIVE

Conclusions

This report presents a disciplined planning process for identifying long-term R&D needs and opportunities that are directly related to high-level R&D needs. It is a top-down approach that begins with strategic analyses for the purpose of developing high-level R&D needs, transitions to system evaluations to establish potential roles for advanced technologies and to assess development uncertainties, and concludes with technology evaluations to identify fundamental problems and R&D solutions.

Developing High-Level R&D Needs: Steps 1 and 2

The approach developed in Chapter Two provides a means of establishing long-term energy R&D needs while recognizing the inability to predict the future. The 2050 threat matrix illustrates the problems that threaten a steady supply of affordable energy. Moreover, the matrix provides a means of prioritizing the problem areas. In the example threat matrix developed for this report, all significant long-term problems associated with current energy-use patterns were identified and were then separated into three groups to focus attention on those judged most serious.

The threat matrix approach appears to have flexibility and applicability beyond what was exercised in this study. In particular, threat matrices may be useful tools for establishing nearer-term energy R&D needs. In looking at multiple time frames, one should anticipate a shifting of the emphasis placed on various problems. For example, global climate change does not pose a near-term threat to affordable supplies of energy, whereas air pollution is severely restricting energy options in certain metropolitan areas and would almost surely be listed as a serious problem in a near-term energy threat matrix.

In our 2050 threat matrix, potential problem areas were separated into three groups: (1) no or negligible problems, (2) problems that are unlikely to cause a major adverse impact on national energy supply and use or that can be addressed without high costs, (3) and serious problems that may severely limit future energy options. The threat matrix approach can support a finer division of problem severity, including the assignment of numerical values. It is not evident, however, that finer detail would provide any advantage in establishing R&D needs.

The threat matrix approach should also be applicable to other federal research programs aimed at technology development. In contrast to private firms, in which profit is the motive, many government organizations have missions whose objective is to ensure that one or more public needs will be met. Implicit in such missions is the existence of threats, and the threat matrix approach may be useful for delineating problem areas and establishing needs for both R&D and other program areas.

In this study, long-term energy R&D needs are purposely articulated at a fairly high level. Our principal objective was to include all possible approaches and concepts for addressing the problem areas designated as most serious. The next steps in the IntraTech approach expose the performance risks associated with such concepts.

Evaluating Technologies: Steps 3 Through 6

The case study showed that the IntraTech approach can identify the long-term energy needs that a technology concept addresses, identify the key performance factors and risks, identify R&D needs and opportunities, and provide insights regarding alternative development options.

The case study—power generation based on coal gasification—was purposely selected so that it fell completely in one of the three product areas, electric power. However, many technologies under development within DOE cut across product lines. For example, if coal gasification were selected as the case-study topic, all three product areas would be relevant: electric power via IGCC, processed liquid fuels via indirect liquefaction of coal, and, obviously, gaseous fuels. We recommend selecting a technology area so that it fits within a product area. In some cases, this means that more than one IntraTech analysis will be needed. The fact that a particular technology area supports multiple products (or goals) raises the potential value of research opportunities relevant to that technology. While not developed within this study, this issue is an important topic for developing and prioritizing the overall long-range R&D portfolio.

An advantage, as well as a limitation, of the IntraTech approach is that it stays within a technology area. The method compares the prospective performance and costs of an advanced technology to what is commercially available. As such, this approach may be especially useful to DOE officials responsible for planning, justifying, and implementing long-term R&D programs directed at specific technology concepts or groups of related technology concepts.

At its current state of development, the IntraTech approach is not appropriate for portfolio management. This limitation became evident during our analysis of program balance within the DOE solid oxide fuel cell program. Whether the IntraTech approach can evolve into an InterTech approach suitable for portfolio management remains an open issue. Meanwhile, IntraTech can be used to provide senior-level federal officials with a consistent and comparable set of information on technologies of interest that might better inform the current intuitive/consensus-based process of portfolio management and budget allocation.

Implementing the IntraTech Approach

Like any planning tool, the IntraTech approach will be of value only if it is used. One approach is for DOE or one of its offices to conduct a beta test of the IntraTech planning method. For example, a "Red Team" might use it in the annual planning and budgeting cycle, in parallel with the usual DOE process. Such a test would show whether implementing the IntraTech approach would bring DOE to a different set of conclusions than those resulting from the standard planning process. The evaluation of the beta test would have three components: (1) an evaluation of the trial itself, (2) a review of experience-based modifications to the IntraTech process, and (3) if appropriate, an adoption plan for use by DOE.

APPENDIX A

Threat Matrix Analyses

This appendix summarizes the analyses on which the threat matrix presented in Chapter Two is based.

Supply Constraints

Estimates of U.S. coal resources indicate that low-cost coal should be abundant in 2050. Recoverable coal reserves are estimated at over 250 billion tons (Energy Information Administration, 1995), and at current production levels, only one-fifth of these demonstrated reserves will be produced by 2050.

The estimated technically recoverable amount of petroleum (crude oil and natural gas liquids) left in the United States is less than 200 billion barrels. Domestic annual production peaked in 1970 at 4.1 billion barrels and has been dropping steadily since 1985. Current annual production is 2.8 billion barrels (Energy Information Administration, 2002b). At this production rate, more than two-thirds of the estimated recoverable resource base currently in place will be depleted over the next 50 years.

Current estimates place the amount of technically recoverable natural gas in the United States at about 1500 trillion cubic feet (Energy Information Administration, 2002c). At the current annual production rate of 20 trillion cubic feet (Energy Information Administration, 2002b), two-thirds of the remaining natural gas resource will be consumed by 2050. Moreover, most energy projections show increased domestic natural gas production over the coming 25 years. For example, the "reference case" EIA projections show production surpassing 27 trillion cubic feet in 2025 (Energy Information Administration, 2003a). Even higher production rates could be seen if natural gas is used to displace coal in order to reduce carbon emissions in the production of electric power.

There is a significant chance that affordable domestic supplies of both petroleum and natural gas will not be available in 2050. The affordability of imported petroleum and natural gas will hinge on the amount of new discoveries and the extent to which international demand for these fuels increases, especially in developing countries. For these reasons, supply constraints are indicated in the threat matrix (Figure 2.1) as potentially very serious problems affecting petroleum and natural gas over the long term.

A massive international shift to nuclear power by 2050 could be constrained by limited availability of uranium ore and ore-processing facilities. During the 1970s and 1980s, this limitation motivated research in advanced nuclear reactors capable of breeding additional fissionable fuel. Currently, there is an excess of nuclear fuel on the market, prices for uranium ore and enriched-uranium fuels are low, and there is little motivation to invest in uranium resource development or fuel-production facilities. This situation has arisen because the use of nuclear power has grown more slowly than anticipated. For example, no new nuclear power plants are on order or under construction in North America, South America, or Western Europe (Energy Information Administration, 2002a).

A 2001 study by the International Atomic Energy Agency warns of possible shortfalls in the uranium resource base and in ore processing by or before 2050, under the assumption that the use of nuclear power in 2050 will be at least three times greater than current levels. If nuclear power is to have a significant future role in world and U.S. electricity supply, this is a fairly modest assumption. More problematic are estimates of the international uranium resource base. Considering the experience in petroleum and natural gas, we should anticipate that increased uranium fuel prices will motivate additional exploration that will lead to a major increase in the size of the international uranium resource base. Because this issue is not resolved, we label it as moderately important in the threat matrix.

Currently, hydroelectric generating stations supply about 10 percent of annual electricity generation in the United States.[1] Most of the economically attractive sites for hydroelectric power are already in use. Lack of economic and environmentally acceptable sites is likely to result in, at best, limited growth in the use of hydropower to generate electricity.

National Security

Most of the world's petroleum resources are concentrated in countries that border on the Persian Gulf. Continued access to these resources could require armed intervention and the commitment of military and defense expenditures required to support such intervention. Additionally, diplomatic requirements for continued access could compromise other national security objectives.

Greatly increased U.S. reliance on nuclear energy is likely to be accompanied by increased global reliance. To meet a significant fraction of the projected international demand for electric power, hundreds of nuclear power plants will have to be built, many in countries with unstable governments. This raises the very serious threat that fissionable materials could be diverted from peaceful purposes to nuclear weapons development.

Air Pollution

State-of-the-art environmental control technology can greatly reduce airborne emissions of pollutants that have been the primary targets for control in new coal-fired electric generation plants. These pollutants include sulfur oxides, nitrogen oxides, and particulates. Available control technologies include advanced flue-gas desulfurization systems, staged combustion combined with catalytic nitrogen oxide reduction, and advanced filter systems for particulate control. While these high-performance environmental control systems are not being deployed in current plants because of limited cost and operating information, their use in newly constructed plants is unlikely to result in significant increases in capital or operating

---

[1] Record droughts in 2000 and 2001 caused the portion of electricity generated by hydroelectric stations to drop from 10 percent to 6 percent.

costs (i.e., to increase them by as much as 10 percent) as compared with recently constructed coal-fired generating plants. Air pollution is considered to be a potential problem area for coal-fired generating plants in 2050 because of the potential need for extensive control of emissions of trace elements, such as mercury, that are found in coal or trace compounds formed during combustion.

With regard to petroleum and natural gas, air pollution is marked as a potential problem because of the emissions resulting when these fuels are used in transportation and small-scale applications. Over the near and mid-term, this may be one of the more serious threats associated with energy use, but current and emerging technology appears capable of supporting numerous approaches that can address these emissions. For this reason, air pollution from oil and natural gas is not designated as a serious threat on the long-range threat matrix.

Land and Water

Long-term coal use is potentially threatened because of the effects of mining on land (e.g., changing natural contours and subsidence) and water resources (e.g., from acid mine drainage, diversion of streams, disposal of solid wastes).

Coal-mining operations use heavy equipment and usually modify land contours. There are reclamation laws in effect for surface mining operations, but the final land contours may differ from pre-mining conditions. Contour modifications, solid-fill disposal, and acid mine drainage have the potential to seriously harm water resources.

As currently practiced, coal use generally involves on-site disposal of ash and other solids resulting from coal combustion and associated environmental control systems. Adverse impacts on land and water resources may prohibit coal mining or power plant siting in specific areas. Overall, these issues may lead to slightly higher costs; however, it is highly unlikely that protection of land and water resources constitutes a major threat to long-term coal use.

Oil and gas extraction and transport have the potential to degrade ecosystems—i.e., complex associations of vegetation and wildlife—in areas affected by extraction activities, pipelines, and tanker routes. Of particular concern is loss or disruption of habitat of species identified as sensitive to ecological changes. Additionally, surface water and groundwater can be adversely affected by accessing drilling areas, staging equipment, setting up rigs, drilling, and operating wells (LaTourrette et al., 2003). The possible damage to ecosystems and water may preclude extraction in certain areas, but it is unlikely to be a major long-term threat to petroleum and gas extraction within the United States. A variety of technical methods are available to mitigate or significantly reduce the risk of environmental damage. While mitigation measures might increase production costs, the significant driver for cost increases is the depletion of low-cost oil and gas resources.[2] Therefore, land and water degradation is not designated in the threat matrix as a serious potential threat to continued reliance on oil and natural gas.

---

[2] The possible exception may be coal-bed methane development, which generates large volumes of potentially hazardous formation fluids.

Land and water issues are marked as a serious threat to reliance on nuclear energy because of potential problems associated with nuclear-waste management, especially if a massive shift were made to nuclear fission-based electricity generation. Experience to date with the siting of a high-level waste repository proposed at Yucca Mountain, Nevada—on the edge of the already contaminated Nevada nuclear weapons test site—indicates that the siting process for waste disposal can be long and contentious. No other nation has yet succeeded in siting and emplacing spent fuel in a permanent repository.

Disturbance to land and adverse ecological impacts, especially to fish, preclude construction of significant additions to the U.S. hydroelectric power inventory.

Climate Change

Certain gases in the earth's atmosphere play an important role in controlling surface temperatures and climate. In essence, these gases absorb some of the heat radiating from the earth's surface and contribute to maintaining higher surface temperatures than would otherwise occur. Since this process is similar to the role of glass in a greenhouse, these gases are collectively termed *greenhouse gases*. Water and carbon dioxide are the most abundant and, for that reason, the most influential of the greenhouse gases. There are many other greenhouse gases in the atmosphere, the most important of which for controlling climate appear to be methane and nitrous oxide.

The major greenhouse gases occur naturally. Since the industrial revolution, appreciable amounts have been added by human activities, especially the combustion of fossil fuels, which releases carbon dioxide into the atmosphere. A fairly strong scientific consensus is emerging that anthropogenic emissions of greenhouse gases will cause changes in the earth's climate. More uncertain is the relationship between specific global carbon dioxide emission levels and the impacts on global and regional climate. Also uncertain are the measures that nations will take to reduce greenhouse gas emissions. Consequently, it is not possible to predict U.S. or global emissions of carbon dioxide 50 or 100 years into the future.

The problem of global climate change could have a major impact on the use of all fossil fuels. For this reason, climate change is listed in the threat matrix as posing a very serious threat to continued reliance on all fossil fuels.

All fossil fuels yield carbon dioxide on combustion. Coal yields the greatest amount of carbon dioxide per unit of energy released, while natural gas yields the least. The difference in carbon dioxide emissions is the result of the carbon-to-hydrogen ratio in the various fossil fuels. Coal has a carbon-to-hydrogen ratio about twice that of crude oil and about four times that of natural gas. The carbon dioxide coefficients (kg of carbon dioxide released per mmBtu of heating value) of the major fossil fuel resources are shown in Table A.1.

Table A.1
Carbon Dioxide Emission Coefficients of Major Fossil Fuels

Fuel	Carbon Dioxide Coefficient (kg/mmBtu)
Coal	95
Crude oil	74
Natural gas	53

Safety

Three liquefied natural gas (LNG) terminals are presently in operation in the United States. Two of them (Everett, MA, and Lake Charles, LA) receive imported gas, while the third is an export terminal located in Alaska. An LNG ship holds between 70,000 and 200,000 cubic meters of LNG. On vaporization, each cubic meter of LNG produces about 625 cubic meters of natural gas. To date, there have been no LNG spills approaching the amount held in a modern tanker or terminal. Smaller experiments indicate that a spill can stay at ground level and can quickly spread.

In 1944, a small LNG storage tank (about 6000 cubic meters) in Cleveland, OH, failed. The contents moved into the local sewer system, and 128 persons were killed when the gas ignited. A recent calculation predicted that an instantaneous release of 20,000 cubic meters could result in a flammable mixture that could travel up to 3.3 miles in 25 minutes (Federal Energy Regulatory Commission, 2001). It is also possible that a massive explosion could occur under certain conditions in which ignition would be delayed.

For these reasons, safety considerations pose a threat to greater reliance on natural gas if such reliance requires LNG imports appreciably above present levels of less than 1 percent of natural gas consumption. A number of technical and operational measures[3] could be implemented to address this threat. Consequently, LNG safety is not designated as a major long-term problem in the 2050 threat matrix.

Growth in the use of nuclear energy may continue to be stymied by concerns over large accidental releases of long-lived radioactive materials at power plant sites, the transport of nuclear waste from reactor sites to a repository, and the long-term (up to 10,000 years, according to current regulatory standards) health implications of radionuclides migrating from a repository. Since the accidental release at the Three Mile Island nuclear power plant, the nuclear industry appears to have successfully implemented procedures for preventing health-threatening accidental releases at power plant sites. Moreover, advanced power plant concepts that could preclude runaway reactions are under development. Relatively little research has been conducted over the past decade on transport risks, although experience during this period suggests that public health and safety can be adequately protected. The health implications of exposure to radionuclides released from a repository depend on numerous assumptions about the demographics and lifestyles of potentially exposed populations.

---

[3] Operational measures include siting large LNG storage facilities in sparsely populated areas, constructing offshore unloading terminals, and providing increased protection from terrorist attacks.

APPENDIX B

Current Coal-Fired Power Generation

The current state of the art in coal-fired electric power generation involves the combustion of pulverized coal in a boiler. Radiant heat released during combustion raises high-pressure steam. The temperature of the steam is further raised (to about 540°C) by heat exchanged from the hot flue gases. The high-pressure steam is sent to a turbine, which is directly connected to an electric generator.

As shown in the threat matrix, power generation from coal raises long-term concerns in three areas: air pollution, degradation of land and water resources, and global climate change.

Historically, the major problem of coal use for power generation has been air pollution, in particular, emissions of sulfur oxides, nitrogen oxides, and particulate matter. Today, however, reliable, efficient, and affordable technology exists that can significantly reduce emissions of these three groups of pollutants. These emissions are unlikely to be a major concern in future power plants built using new, proven environmental control technology.[1] But another issue has arisen: Mercury, which is generally recognized as highly hazardous to human health, is a trace impurity in coal, and coal-fired boilers are estimated to emit nearly half of the nationwide airborne emissions of mercury (Keating, 1997). Technology appears to be available for controlling mercury emissions from pulverized-coal-fired power plants, although significant technical issues remain unresolved, including the impact of fuel variability on mercury controllability (ENSR, 2003). Congress is currently considering legislation that would impose caps on airborne mercury emissions associated with power generation.

The environmental impact of coal mining operations on land and water resources is summarized in Appendix A. While mining may be hindered or prohibited in particular areas because of adverse environmental impacts, it appears unlikely that they represent a major barrier to continued coal use over the long term.

Coal contains many impurities, most of which are contained in the ash residue, which generally constitutes about 10 percent of the weight of the coal delivered to a power plant. This ash residue is stored at power plant sites, where measures are taken to prevent leaching, which can degrade groundwater and streams. Environmental control systems generate additional solid wastes that are stored at power plant sites. The disposition of solid wastes at power plant sites is an important issue. On-site solids disposal requires a large and unsightly plant site, which could seriously impede permitting of future coal-fired power

---

[1] Sulfur and nitrogen oxide emissions constitute a serious problem in many existing coal-fired power plants. In particular, only 27 percent of existing coal-fired generating capacity is from power plants fitted with sulfur oxide removal systems (Energy Information Administration, 2002b).

plants. Over the long term, it is likely that reasonable-cost approaches can be developed and implemented that address the on-site solids-disposal problem.

The major threat to continued coal use is global climate change. Coal-fired power plants produce about 50 percent of the U.S. emissions of carbon dioxide, one of the principal chemical species known to be capable of causing global climate change. About one-third of U.S. emissions of carbon dioxide come from coal-fired power plants (Energy Information Administration, 2002d). This is a significant amount, considering that electricity generated from coal constitutes less than 9 percent of the energy delivered to end-users. This disproportionate release of emissions results from two factors. First, due to its chemical composition, coal has the highest carbon dioxide emission coefficient (see Table A.1 in Appendix A) of any fossil fuel. Consequently, for the same amount of energy yield, coal combustion produces more carbon dioxide than does combustion of other fossil fuels. If abundant supplies of low-cost natural gas were available, using natural gas rather than coal in existing boilers would decrease carbon dioxide emissions by about 40 percent.[2]

The second factor that drives high carbon dioxide emissions from coal-fired power plants is the low energy efficiency of these plants. In 2001, the average energy efficiency of all coal-fired power plants operating in the United States was only 33 percent. Basically, two-thirds of the energy input was lost as waste heat. Recently built coal-fired power plants are designed to operate at efficiencies of about 36 percent. The state of the art is between 40 and 42 percent, but such highly efficient plants do not appear to be economical in the United States, given relatively low prices of delivered coal.

When the energy efficiency of power plants is increased, less coal is required to meet the same power demand. This translates into lower carbon dioxide emissions as well as across-the-board environmental benefits such as less disturbance to land from mining, less coal transported, and reduced solid wastes. Figure B.1 (on the following page) illustrates how carbon dioxide emissions decrease when efficiency is increased above 33 percent. In particular, pulverized coal power plants operating at state-of-the-art energy efficiency would reduce coal use and carbon dioxide emissions by about 20 percent as compared to the current mix of coal-fired power plants now operating in the United States.

[2] Even greater reductions would result if new, highly efficient natural gas-fired combined-cycle plants replaced existing coal-fired plants. In particular, state-of-the-art commercial natural gas-fired power plants operate at energy efficiencies (HHV) exceeding 56 percent.

Figure B.1
Effect of Increasing Energy Efficiency on Carbon Dioxide Emissions

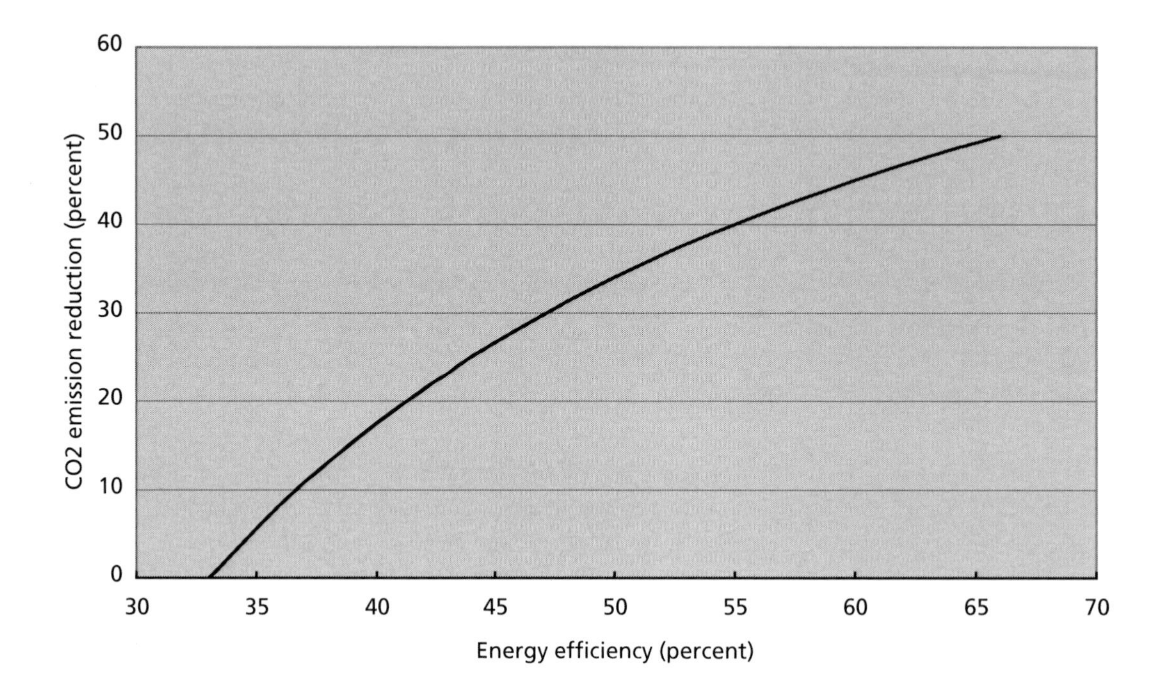

References

Buchanan, T., et al., Parsons Infrastructure & Technology Group, Inc., *Updated Cost and Performance Estimates for Fossil Fuel Power Plants with CO_2 Removal,* Electric Power Research Institute, Interim Report, 1004483, December 2002.

Energy Information Administration, *Annual Energy Outlook 2003*, U.S. Department of Energy, DOE/EIA-0383(2003), January 2003a.

Energy Information Administration, *Annual Energy Review 2001,* U.S. Department of Energy, DOE/EIA-0384(2001), November 2002b.

Energy Information Administration, *Coal Data: A Reference*, U.S. Department of Energy, DOE/EIA-0064(1993), February 1995.

Energy Information Administration, *Emissions of Greenhouse Gases in the United States 2001,* U.S. Department of Energy, DOE/EIA-0573(2001), December 2002d.

Energy Information Administration, *International Energy Outlook 2002,* U.S. Department of Energy, DOE/EIA-0484(2002), March 2002a.

Energy Information Administration, *Monthly Energy Review January 2003*, U.S. Department of Energy, DOE/EIA-0035(2003/01), January 2003b.

Energy Information Administration, *U.S. Crude Oil, Natural Gas and Natural Gas Liquids Reserves: 2001 Annual Report*, U.S. Department of Energy, DOE/EIA-0216(2001), November 2002c.

ENSR Corporation, *Multivariable Method to Estimate the Mercury Emissions of the Best-Performing Coal-Fired Utility Units Under the Most Adverse Circumstances Which Can Reasonably Be Expected to Recur,* submitted by West Associates at the Clean Air Act Advisory Committee Mercury MACT Working Group meeting, March 4, 2003.

Federal Energy Regulatory Commission, *Cove Point LNG Project Environmental Assessment*, Docket No. CP01-76-000, July 2001.

International Atomic Energy Agency, *Analysis of Uranium Supply to 2050*, Vienna, 2001.

Keating, Martha H., *Mercury Study Report to Congress, Volume II,* U.S. Environmental Protection Agency, EPA-452/R-97-004, December 1997.

LaTourrette, Tom, et al., *Assessing Natural Gas and Oil Resources: An Example of a New Approach in the Greater Green River Basin*, Santa Monica, CA: RAND Corporation, MR-1683-WFHF, 2003.

Merrow, Edward W., K. E. Phillips, and C. W. Myers, *Understanding Cost Growth and Performance Shortfalls in Pioneer Process Plants*, Santa Monica, CA: RAND Corporation, R-2569-DOE, 1981.

National Energy Policy Development Group, *National Energy Policy*, Office of the President, May 2001.

National Energy Technology Laboratory, *Fuel Cell Handbook, Sixth Edition*, U.S. Department of Energy, DOE/NETL-2002/1110, November 2002.

PCAST 97: President's Committee of Advisors on Science and Technology, *Federal Energy Research and Development for the Challenges of the Twenty-First Century*, Office of Science and Technology Policy, November 1997.

Ratafia-Brown, Jay A., et al., "An Environmental Assessment of IGCC Power Systems," paper presented at the Nineteenth Annual Pittsburgh Coal Conference, September 23–27, 2002a.

Ratafia-Brown, Jay A., et al., *Major Environmental Aspects of Gasification-Based Power Generation Technologies*, U.S. Department of Energy, December 2002b.

Ruth, Lawrence A., "Advanced Clean Coal Technology in the U.S.," paper presented at the 3rd International Workshop on Life Cycle Issues in Advanced Energy Systems, Woburn, England, June 2002.

SEAB 95: Secretary of Energy Advisory Board, *Energy R&D, Shaping Our Nation's Future in a Competitive World*, U.S. Department of Energy, June 1995.

Singhal, S. C., and M. Dokiya (eds.), *Solid Oxide Fuel Cells VI, Proceedings of the Sixth International Symposium*, The Electrochemical Society, Inc., Pennington, NJ, 1999.

United States Department of Energy, *Comprehensive National Energy Strategy,* DOE/S-0124, April 1998.

United States Department of Energy, *The Department of Energy Strategic Plan*, September 30, 2003.

Yokokawa, H., and S. C. Singhal (eds.), *Solid Oxide Fuel Cells VII, Proceedings of the Seventh International Symposium*, The Electrochemical Society, Inc., Pennington, NJ, 2001.